STANDING ALONE

STANDING ALONE

*An American Woman's Struggle
for the Soul of Islam*

ASRA Q. NOMANI

HarperSanFrancisco
A Division of HarperCollinsPublishers

For my mother and my father
And for my son, Shibli
"Ababooboo"

HarperCollins books may be purchased for educational, business, or sales promotional use. For information please write: Special Markets Department, HarperCollins Publishers, 10 East 53rd Street, New York, NY 10022.

HarperCollins Web site: http://www.harpercollins.com
HarperCollins®, ™®, and HarperSanFrancisco™ are
trademarks of HarperCollins Publishers.

FIRST HARPERCOLLINS PAPERBACK EDITION PUBLISHED IN 2006
Designed by Joseph Rutt

Library of Congress Cataloging-in-Publication Data is available.
ISBN 13: 978–0–06–083297–1
ISBN 10: 0–06–083297–5

06 07 08 09 10 RRD(H) 10 9 8 7 6 5 4 3 2

CONTENTS

PART THREE
MAKING THE PILGRIMAGE
February 2003

PART SEVEN
HARVESTING THE FRUITS OF THE PILGRIMAGE
June 2004 to October 2004

PREFACE

This is a tale of a journey into the sacred roots of Islam to try to discover the role of a Muslim woman in the modern global community. This book is a manifesto of the rights of women based on the true faith of Islam. It heralds a revolution in the Muslim world of the twenty-first century.

My book is about sorting out the contradictions about religion. Two defining moments shaped my relationship with my religion: the murder of my friend Daniel Pearl and the birth of my son, Shibli Daneel Nomani. The men who plotted the kidnapping of my friend did the five daily prayers that are one of the five pillars of my religion; and the men who killed my friend did so in the name of my religion. The man who is the father of my baby went to the mosque for his Friday prayers but did not stand beside me when I brought my baby into the world. He considered me illegitimate in the eyes of my religion because, while foolishly in love, we weren't married when we conceived our baby. Others called me a criminal in the name of Islam.

I was very much at odds with my religion. But instead of turning away from Islam, I decided to find out more about my faith. From my home in Morgantown, West Virginia, I embarked on the holy pilgrimage to Mecca, the birthplace of Islam, in the kingdom of Saudi Arabia, one of the most repressive regimes in the world for women. What happened there shocked me. The hajj became the catalyst to my empowerment as a woman in Islam.

The hajj allowed me to walk a living history that pulses with meaning to Muslims. It is best described as a journey that follows the path of Adam, the first prophet in Islam; Abraham, the father of Islam; and Muhammad, the prophet of Islam and the last messenger of God, and a man so revered by Muslims the words "peace and blessings be upon him" follow his name. But I walked with a different mindfulness, following the struggles, strengths, and triumphs of the women of my religion. I traced the path of the great women in Islam starting with Eve, who, contrary to biblical opinion, was not responsible for original sin; Hajar (or Hagar), the single mother of Islam; and Khadijah, the first wife and patroness of the messenger of Islam. As my guides, I had with me living models for the best in Islam, my mother and my father. And as my beacons for the future of Islam, I had my young niece, nephew, and son.

The journey took us to the three holiest mosques in the Muslim world in the cities of Mecca, Medina, and Jerusalem. I crossed thresholds into the space where the Muslim civilization was born. I felt beneath my feet the earth where the first Muslims risked death to practice their religion. And I gazed up at the sky that served as the canopy for the forgotten mother of Islam, Hajar.

On my journey, a remarkable transformation happened in my understanding of my religion. I uncovered the hidden secrets of the strength of Muslim women in the earliest years of Islamic history. I discovered the legacy of Muslim women who marched into battle with spears, challenged the prophet, and sculpted the society that was the first Muslim society. They prayed, worked, and pioneered a new community with men, empowered by their leader, the prophet Muhammad, to contribute fully and express themselves completely. The prophet was indeed the Muslim world's first feminist.

I returned to my home in Morgantown with an awakened understanding of the highest ideals of Islam for both women and men and for the community. The hajj gave me the courage to act on the ultimate conviction that women can be fully engaged members of the Muslim community. What I discovered was sad testimony to the ways in which the original might of women in Islamic history have been erased by religious clerics and men who are more interested in power and control than the ideals originally set out by Muslim society. It led me into a deep confrontation with tradition, power, and fear.

What I realized is that the core values of all religions and cultures are the ones that serve us best: truth, knowledge, love, and courage. Living with *truth* allowed me to free myself from the duplicities, contradictions, and shame that so often constrain us. *Knowledge* about the heart of my religion helped to liberate me as a woman in Islam. Scholars of yesterday and today freed me from oppressive traditions that I was expected to inherit. Through their practice of *ijtihad*, or critical reasoning, and their individual strength in standing up to forces of darkness, they blew open the doors of my religion to me. I am grateful to them for their scholarship. The *love* of my parents, my friends, and my spiritual community encouraged me to explore territories that I had never charted. My mother and my father gave me the freedom to think, explore, and learn. They allowed me to find my voice, the result of which is this book. And they embraced me always with unconditional love.

The final theme of this book is *courage*—the courage to be honest and stand up for justice. It is a theme that is as universal today as it was yesterday and will be tomorrow. I hope to encourage others to live with honesty and to live peacefully with their religion through their own personal searches and faith in the belief that the divine force of the universe is compassionate, loving, kind, and accepting. This is a book about a Muslim woman in America. But this is also a book for women and men of all faiths.

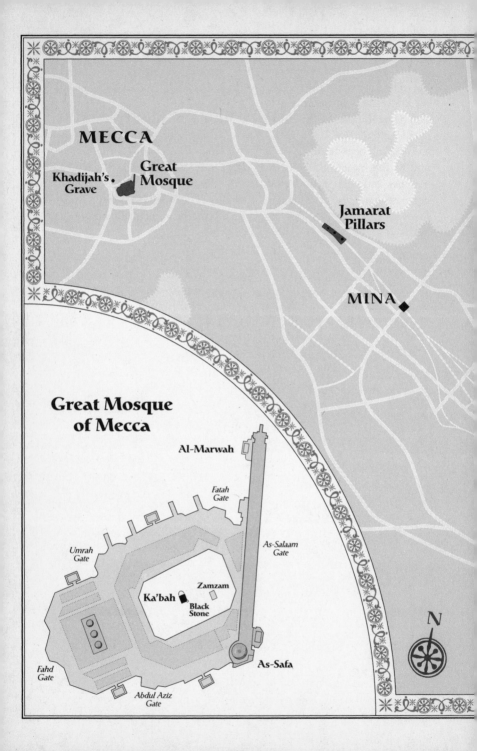

MECCA

Khadijah's
Grave

Great
Mosque

Jamarat
Pillars

MINA

Great Mosque
of Mecca

Al-Marwah

Fatah
Gate

As-Salaam
Gate

Umrah
Gate

Ka'bah

Zamzam

Black
Stone

Fahd
Gate

As-Safa

Abdul Aziz
Gate

N

Medina

Jeddah ◆ ◆ Mecca

*Red
Sea*

**Arabian
Peninsula**

*INDIAN
OCEAN*

250 miles

MUZDALIFAH ◆

Mecca and
the Hajj

1 mile

**Mount of
Mercy**

ARAFAT ◆

PART ONE

EMBARKING ON THE JOURNEY

January 2001 to January 2003

THE DALAI LAMA AND THE SEEDS
OF A PILGRIMAGE

ALLAHABAD, INDIA—One hot winter afternoon, I was lost in India on the banks of the Ganges, a river holy to Hindus. I was meandering with an American Jewish friend on a road called Shankacharaya Marg. By chance, my path intersected with the spiritual leader of Tibetan Buddhists, the Dalai Lama, inside an ashram, and he set me off on my holy pilgrimage to the heart of Islam.

It was January 2001, and I was, quite fittingly, in the city of Allahabad, "the city of Allah," the name by which my Muslim identity taught me to beckon God. In Islam, Allah is our Arabic word for God. Born about a thousand miles westward along the Indian coastline in Bombay, India, I had evoked God with this name from my earliest days.

Although a Buddhist, the Dalai Lama, like millions of Hindu pilgrims, was in a dusty tent village erected outside Allahabad to make a holy pilgrimage to the waters there for the Maha Kumbha Mela, an auspicious Hindu festival. He joined the chanting of a circle of devotees dressed all in white. When they had finished, I followed the Dalai Lama to a press conference in a building surrounded by Indian commandos and his own bodyguards. Religious fundamentalism and fanaticism are wreaking havoc throughout the world, and in India they are redefining Hindu and Muslim communities that used to coexist peacefully. The demolition of a sixteenth-century mosque called Babri Masjid sparked one of India's worst outbreaks of nationwide religious rioting between Hindus and the Muslim minority; two thousand people, mostly Muslims, were killed. The cycle of hatred continued until that day when the general secretary of the sectarian World Hindu Council, Ashok Singhal, called Islam "an aggressive religion." At the press conference an Indian journalist raised his hand. "Are Muslims violent?" he asked.

My stomach tightened. This question reflected a stereotype of the people of my religion, but, alas, the national flag of Saudi Arabia, the country that considered itself the guardian of Islam's holiest cities—two historical sites called Mecca and Medina—includes the sword.

The Dalai Lama smiled. "We are all violent as religions," he said. After pausing, he added, "Even Buddhists."

We all smiled.

"We must stop looking at the past," he continued, "and look at the present and the future."

I sat near the back, my usual spot at press conferences, and pondered his words. I had spent a lot of my life trying to understand my past. My mother and father, Sajida and Zafar Nomani, were children of India when it was still under British colonial rule. I was born in Bombay in 1965, after the country had won liberation. My parents left for America when I was two so my father could earn his PhD at Rutgers University in New Brunswick, New Jersey. My brother, Mustafa, my only sibling, and I stayed with my father's parents until we boarded a TWA jet in 1969 to JFK Airport in New York to be reunited with our parents. I became a journalist, landing my first job at the age of twenty-three with the *Wall Street Journal*. Despite my apparent success, I had difficulty expressing my voice. I wanted to raise my hand, but even though I had been a successful staff reporter for one of the most powerful newspapers in the world for over a decade, I could barely muster the courage to ask questions at press conferences. To justify my fears, I accepted the rationale passed on to me once by a senior journalist at the *Wall Street Journal*: "Don't ask any questions at press conferences," she had told me over the phone as I reported from the scene of a United Airlines plane crash. "That way, nobody will know what you're thinking."

Inviting and beaming, the Dalai Lama triggered something. All of a sudden, I wanted to let others know what I was thinking. I realized I had a responsibility to speak up. As long as I called myself a Muslim, I had to try to bridge the schism between my religion and others. I tentatively raised my hand. To my surprise, the Dalai Lama gestured eagerly at me. I began to speak my thoughts, marking a turning point in my life as I did so.

"Through personal meditation we can transcend ego and power in our own lives," I said. "What is it that our leaders can do to transcend the issues of power that make them turn the people of different religions against each other?"

He looked at me intently and said: "There are three things we must do.

Read the scholars of each other's religions. Talk to the enlightened beings in each other's religions. Finally, do the pilgrimages of each other's religions."

I nodded my head in understanding. I, a daughter of Islam, was in the midst of the Hindu pilgrimage. I had grown up with a mocking understanding of the deities to which Hindus bow their heads, but sitting in a retreat colony amid simple devotees like an elderly Indian Hindu woman named Mrs. Jain, I understood that the spiritual intention of a polytheist is no different from that of a monotheist who prays in a synagogue, church, or mosque.

Just months earlier, I had climbed into the Himalayas at India's border with China and joined about twenty thousand Buddhists in a pilgrimage led by the Dalai Lama. On the last day I had tried to resuscitate an elderly Nepali Sherpa who had gotten caught in a stampede by pilgrims rushing to witness a holy religious sand creation called a mandala. He literally died resting in my hands, and I knew at that moment the universal phenomenon of faith that defines all religions. I had just spent two years speaking to the scholars of the faiths and reading their texts. I had read the teachings of the Buddha. I had read the Bible. I had sat at the feet of a pandit, a Hindu scholar who comes from the upper Brahmin caste of Hinduism. As a woman, I was trying to grasp the role of women in the faiths. I learned that sacred goddesses were integral to early civilizations, such as the Indus civilization from which India sprang.

These societies honored matriarchy and emphasized the power rooted in women. But they mostly evolved into patriarchal cultures in which men are considered more important than women. Most of the principles of goddess worship have disappeared from modern society.

I had not been able to understand my role as a woman in my religion of Islam. When I was a child of seven or eight in Piscataway, New Jersey, I asked my Islamic Sunday school teacher, a kind Egyptian man by the name of Dr. Mahmood Taher, "Why aren't there any women prophets?"

He smiled. "There were," he explained, "*great* women." His lesson to me wasn't in the words that he spoke, but in the kind and open-minded way in which he received my question. He didn't tell me the question was inappropriate. He didn't rebuke me. He took me seriously. With his honest and gentle effort at an answer, he set me on a path of inquiry that I followed into adulthood. In Allahabad, I was still walking that path, eager to find my spiritual home as a woman born into Islam.

I emerged from my thoughts to the sound of journalists pushing back chairs as they got up at the end of the press conference. I jumped to my feet too and slipped to the front to ask the Dalai Lama another question as he left. He paused in front of me. I started to ask him my question. He didn't seem to understand, but it didn't matter. He giggled and lifted my chin with his right hand. It was a gentle and affectionate touch. It felt like a blessing, like a transmission of spiritual power in the greatest tradition of the masters passing their teachings on to students.

I dashed outside to jump into the back of a truck that was following the Dalai Lama's path. It took us to the Ganges River, where I plunged knee-deep into the water as the Dalai Lama, barefoot and laughing, lit candles in a Hindu ritual. He sprinkled himself with water from the Ganges for a centuries-old ritual that Hindus believe washes away their sins so that they can avoid reincarnation. "I'm very happy to be here," the Dalai Lama said, but when asked if he would join the pilgrims bathing in the icy water, he replied, "I don't think so. It's too cold." This attitude reflected a deeper philosophy of the Dalai Lama that I was starting to appreciate. At the press conference, he said, "I always believe it's safer and better and reasonable to keep one's own tradition or belief." It was dark around me— the day was slipping into night—but at that moment a light went on inside of me. I understood what the Dalai Lama's words meant to me. I had done the Buddhist pilgrimage. I was doing the Hindu pilgrimage. I had never done my own pilgrimage—the pilgrimage to Mecca called the hajj. I formed an intention, at that moment, to do my pilgrimage.

Mecca is like the Vatican for Catholics or Jerusalem for Jews. It is the center of our religion. It sits in a vast space of desert called Arabia between the Red Sea and the Indian Ocean. The Qur'an is the Bible of Islam, literally meaning "recitation," and the Qur'an says that it was near Mecca where Adam and Eve—revered in Islam as they are in the other world religions—were reunited after they descended to earth from the heavens. It is where Islam says Abraham, considered the father of Jews, Christians, and Muslims, abandoned a woman named Hajar to live with their son Ishmael. In the Bible and in Jewish history, she is known as Hagar. Mecca is where the Islamic historians say Ishmael married a bride from a tribe called Jurhum; from it sprang the tribe of Quraysh. Centuries later a man named Muhammad was born in Mecca into the tribe of Quraysh in the seventh century A.D. He became the prophet for the religion that is Islam. In Mecca, he started hearing revelations from God that became the Qur'an, making the city the birthplace of Islam.

The Qur'an says that it is the duty of all able-bodied Muslims to do the pilgrimage to Mecca, but I had never even thought about going. In fact, it's a pilgrimage that most Muslims can never take. The hajj is no simple journey. It is an arduous spiritual and physical rite that lasts only five days during a month on the lunar Muslim calendar called Dhul Hijjah, or "the month of the hajj," in Arabic. The hajj is meant to be a time to absorb the central messages of Islam: that Islam means having a special relationship with God based on surrendering to divine will and praying to and revering God; that there is a kinship among people that expresses itself through sacrifice for the benefit of others; that life is about struggle—a battle to secure a livelihood and ensure that good triumphs over evil.

For women, the hajj is given the value of struggle, or jihad. The concept is daunting. Jihad is normally associated with military combat, but its deeper meaning is a struggle within our souls to live by the highest spiritual principles we can embrace.

It's said that the prophet Muhammad's wife Aisha asked him, "Do women have to make jihad?"

The prophet replied, "Yes, the hajj and *umrah*." Umrah is an off-season pilgrimage that happens anytime other than the five designated days of hajj.

I knew that this jihad beckoned me, and the idea of journey felt familiar to me. My family had been early pilgrims of another sort when my family migrated to America. In Arabic, migrants are called *muhajir*, a word that coincidentally sounds like *hajj*. From my earliest days, I had been a person on pilgrimage.

REFLECTIONS ON LIFE AS A DAUGHTER OF ISLAM

The problems with the Christians start, said Father, as with women, when the hudud, or sacred frontier, is not respected. . . . To be a Muslim was to respect the hudud. And for a child, to respect the hudud was to obey.

Fatima Mernissi (Moroccan scholar),
Dreams of Trespass: Tales of a Harem Girlhood (1994)

HYDERABAD, INDIA—I stared at the bare-chested laborer standing on the roof of my childhood home in India, hammering at its final remnants. It made me reflect on my roots and the imprint they left on my identity.

When I left India at the age of four, my grandmother—whom I called Dadi, meaning "paternal grandmother" in my native language of Urdu—dressed my brother and me in matching outfits cut from the same striped cloth (in case we got separated and had to be matched), and she lined my

eyes with black kohl, or eyeliner, to protect me from the evil eye. It served
me well in a life that, much like most lives, has encountered tests. We
lived first in Piscataway, New Jersey, where I spent my girlhood trying to
find a place for myself as an immigrant child of America. Watching a children's TV program called *Romper Room*, I waited for the hostess to call
out my name when she greeted children in TV land called Mary, Sue, and
John, but I never heard her say my name. At home I grew up following
the script of a traditional Muslim girl. I stopped wearing dresses when I
was nine years old because my father, with his traditional Muslim sensibilities, felt that it wasn't proper for me to keep baring my legs.

When I was ten, we moved to Morgantown, West Virginia, where my
father got a job as an assistant professor of nutrition at the local university,
West Virginia University. Morgantown is tucked into a north-central
corner of the state about seventy-five miles south of Pittsburgh and two
hundred miles west of Washington, D.C., along a river called the Monongahela. About 90 percent of the population is white. West Virginia University is known for its football team, the gold-and-blue Mountaineers, but
it has also churned out a record number of Rhodes scholars to Oxford University. West Virginians have a fierce mountain tradition of independence
that I seemed to absorb. I was always proud to be on my side of the boundary line between Virginia and West Virginia. When the war over slavery
broke out, our territory had stood on the right side of the issue, breaking off
from Virginia to create the free state of West Virginia. In Morgantown, we
had a dynamic intellectual community. West Virginia University was the
richest academic enclave in the state. Drawn by the university, immigrants
from India were one of its largest minority groups, next to immigrants from
China. I went to sixth grade at Evansdale Elementary School, across University Avenue from my family's simple three-bedroom apartment in the
WVU faculty housing. Unknown to me, feminist scholarship was just
starting to take root on campus, where Dr. Judith Stitzel, the mother of
one of my classmates, was starting to teach women's studies courses. By no
coincidence, her son, David, took my arm and led me across one of the
greatest divides that defines traditional Islam, and most of our world—the
divide between males and females. He was my square dancing partner. Mrs.
Gallagher, our sixth-grade teacher, had sent a note home with all of her
students, asking our parents' permission to let us learn to square-dance. My
mother invoked what she had been taught in her Muslim family against
boys and girls dancing, but I begged and begged for permission to square-
dance. Finally, my mother relented.

Some might say that was when my troubles began. But for eight years I lived by most of the *hudud*, or rules of my Muslim culture. I didn't protest when I had to sit with the women in the kitchen while the men sat on the nice Montgomery Ward living room sofas. I could hear the roar of my father's voice as the men engaged in political debate. As I grew up, I cared about the civil war in Lebanon and the Iranian hostage crisis. But I never felt I could enter the men's space, and I didn't—except to whisper messages from my mother to my father to stop talking so loudly.

I knew enough, though, to recognize that women were restrained just because of the gender into which we were born. My junior high journal for Mrs. Wendy Alke's English class is filled with snapshots that reveal that it was in my character to be a free spirit. I chronicled the biking accidents, the kickball games, and the other adventures that filled my free time. Not long after I moved to Morgantown, I shared a seat on my bike with a friend. "The handlebars started shaking. I was tense when all of a sudden the bike went down! We both fell, and I got most of the impact since I was up front! We had to walk the bike all the way back to the Med Center Apts., and on our way we saw a car and thought wow! If that had been two minutes later, we could have been run over!" Another time, I recounted how I broke my arm jumping off a wooden fence in my rush to play baseball with my brother and his friends. "I was going to play baseball. The log twisted, and I lurched forward. I got up and oh! yelped in pain. My arm had been broken, and I walked home with the help of my friend." These would have been ordinary childhood stories except that in my life they were also symbolic of the freedoms my parents allowed me as a girl. In traditional Muslim cultures around the world, girls aren't allowed to ride bikes in public; they aren't allowed to play baseball with their brothers; and they most certainly aren't allowed to walk home alone. I started earning money before I hit my teen years, babysitting neighborhood children named Bobby and Misty. I chronicled the night I earned $2.50. This was also remarkable because it set me on a path toward economic independence that so many women in more traditional Muslim culture aren't allowed.

It is clear from my childhood expressions that I looked to God for help in my life. As it sleeted outside one November day, I wrote that the Condors won their kickball game during lunch that day. "That means we are tied with them for 1st place and have to play them to see who is No. 1." Invoking a Muslim phrase that means "God willing," I wrote, "*Inshallah*, we are." When my brother fell ill one summer, I took the blame. Earlier I had gotten jealous that he was healthy while I was sick, and I yelled, "I wish you would die!" With my brother sick, "I started crying and crying and everyone else tried to hold it in. We were all praying and praying!" When my brother survived, I prayed in relief and vowed never to curse anyone. "I was afraid we would lose bhaya [the Urdu honorific for "older brother"]! Thank you so much God for teaching me not to say such bad things and for saving bhaya!" So much is said about Catholic and Jewish guilt, but Muslim culture has its own guilt trips, and I absorbed all the messages that told me my sins could cause damnation. To counter these messages, I looked for inspiration in other sources.

From early on I found strength in the stories of women who challenged tradition. I talked in my journal about Louisa May Alcott's *Little Women*, a tale of strong women, as "my most favorite book."

At many different times in my life, I also felt my culture trying to confine me and define me. From that early age, I could feel the difference between circumstances that were oppressive and those that weren't. I enjoyed a gathering one night celebrating the Hindu holiday of Diwali, or a festival of lights. "Us girls had relay races in the hall and arm wrestling (I beat them all). . . . It was fun all in all." I continued: "The next night . . . there was an Islamic association party. It stunk! The ladies had to go up to a little efficiency apt. (owned by one of the members) because they weren't to sit with men. There were like 15 people in one dinky room! The men carried the food up and oh! it was as if we were in jail!"

As I entered into adulthood I began confronting the boundaries in my life, accepting them at times and daring to challenge them at other times. My father had his own struggles reconciling his culture with his beliefs, but as a scientist he firmly believed in having an open mind and pursuing intellectual inquiry, and he encouraged me to develop these attributes. My father crossed state borders to drive me to New York City so that I could do a summer internship at *Harper's* magazine, but he was also crossing a much more profound kind of line: the cultural tradition that a daughter didn't leave her father's home except to go to her husband's house.

Indeed, to respect these traditions, my parents told me to apply to only my hometown school of West Virginia University, but even there I continued resisting traditional Muslim boundaries. At the age of eighteen, I kissed a man for the first time, and he wasn't my husband. In the study carrels in a building called Colson Hall, our shoes slipped off during an all-night study session, his toes crossed the unspoken physical boundary that my culture and religion had put around me, and he dared to touch my bare feet. The next year I crossed the most sacred of boundaries of a woman's body and consummated my love, but it wasn't my wedding night. I wept in confusion over the truths of my physical and emotional urgings and the expectations of my religion and tradition.

I broke my parents' hearts with my social trespasses. I tried to live a double life, but they knew enough to be disappointed. Still, my parents did not remain captive to their cultural traditions, because higher values overrode their fears, and they allowed me to do my graduate work at American University in Washington, D.C. By doing so they helped me find economic opportunity and professional status. I worked for twelve years as a journalist for the *Wall Street Journal* beginning in 1988, flying into new cities, diving into rental cars, and navigating my way to interview CEOs and senators. I spent my young adulthood trying to understand the amalgamation of identities within me.

In 2000 I took a leave from the *Journal* and traveled alone to India to report and write a book. If my Indian world is divided into a "North" that includes the West and a "South" that includes the East, I am a daughter of the South, but a woman of the North. I went to India as an author to research a book on Tantra, an ancient Hindu philosophy in which feminine powers and sexuality are a critical part of worship. I had written a front-page article for the *Wall Street Journal* about the big business of selling Tantric concepts of sacred sexuality at weekend workshops from Santa Cruz, California, to Ottawa, Canada. I had thought I would wander the caves of India studying with Tantric masters, but my itinerary soon became a journey into the corners of my own identity as I tried to traverse the dualities in my life. I thought I was searching for love, but I was in fact searching for the answer to the question of who I was as a woman.

As I traveled in India I embodied the values of self-determination that I had learned in America. To be mobile, I dared what had been unthinkable to me even in America: I learned how to ride a motorcycle. It was a scooter by U.S. standards—a sleek, black, 100-cubic-centimeter machine—but it was my vehicle of empowerment. I rode that Hero Honda Splendor into the Himalayas, having cut my long hair and wearing pants and jackets to resemble a man. Women didn't ride motorcycles there. But no matter how high I went into the Himalayas or how far away from home I traveled, the voices of traditional values echoed within me.

ISLAMIC RED TAPE

LUCKNOW, INDIA—Following my encounter with the Dalai Lama, I drove to visit my elderly aunt, Rashida Khala (*khala* means "maternal aunt"), in the city of Lucknow. I figured I could do the hajj on my way back to the United States. As it turned out, the closest I could get to Mecca was walking into the Saudi Arabian Airlines office, near the bustling Lucknow neighborhood of Hazratganj.

At first, a man named Nadeem told me there would be only an extra $75 stopover fee on a ticket to New York to do a trip to Mecca.

Awesome, I thought. I am a visual learner: until I go someplace, I don't know where it is on the map. I had no clue where Saudi Arabia was located, beyond knowing it was somewhere in the Middle East, and it was a relief to know it was somewhere along the flight path between the United States and India. Saudi Arabia is a short flight westward from New Delhi. Then the ticket agent added a caveat: "You must go with your *mahram*."

"My what?"

"Your mahram. Your father, your husband, your son, or your brother."

"What?" I asked. Alone, I had jetted into the cities of the world—among them, Bangkok, Delhi, Tokyo, and Paris—but I couldn't fly into the holy cities of my religion? He explained: sharia, or Islamic law, in Saudi Arabia rules that a Muslim woman must do hajj with a mahram, either a husband or an adult male escort who can't legally marry her—her father, son, or brother. Uncles and cousins do not qualify.

Historical anecdotes from the time of the prophet Muhammad are used to support this rule. It's said that a man came to the prophet Muhammad and said, "My name has been included in jihad and my wife has left for the hajj pilgrimage."

The prophet replied: "Go and perform the hajj with your wife."

To me, that seemed interesting, but it certainly didn't make it a rule. For all that is written in the Qur'an about the hajj, no mention is made of chaperones being required. I had only the Saudi Arabian Airlines ticket officer to ask my questions of "Can I go with a tour group?"

"Perhaps, but you must come and go with them. The leader has responsibility for you."

Responsibility for me?

I was thirty-five years old. I had been independent and self-sufficient since the *Wall Street Journal* started cutting me paychecks. I'd driven a motorcycle through the Himalayas. I'd interviewed President Bill Clinton in the Rose Garden of the White House. Surely, I could take care of myself. My thoughts were interrupted by my physical mirror image—a woman walked inside the office shrouded in a black burka, a kind of graduation gown with a loose ninja hood; only her eyes were visible. A man came in with her. She flipped open her passport. The picture inside showed her face, but her hair was still completely shrouded. I wondered what identity I would have to assume to make this journey. Did I have to become *her*? I cherished cousins and aunts who dressed in burka. But as much as I loved and respected them, I didn't think I'd ever adopt their external persona. I left frustrated.

In the house of my elderly aunt, Rashida Khala, a doting, pious, and beautiful woman, I met a distant cousin. He was planning on going on the hajj with his wife. "Can I go with you?" I asked eagerly.

"We shall see," he answered.

He was my elder. Surely he could qualify as my mahram. Surely the Saudi government didn't care if he was a distant relative. Surely, it turned out, it did. I rode my motorcycle for almost seven hours (with a rest overnight) to my ancestral village of Jaigahan, where women don't emerge from the cloisters of their houses except in burka. Yet I couldn't travel to Saudi Arabia with a cousin. I had to be accompanied by a direct bloodline male relative. On top of that restriction, the fundamentalist Hindu government in power in India didn't seem to rank Muslim pilgrims as a top priority. Fundamentalism in Hinduism was like fundamentalism in all religions: its adherents believed in the supremacy of their faith over others. Its politicians wanted to turn India's secular state into a Hindu nation. This was an issue for hajj pilgrims in India, where, as in many countries outside the Western world, pilgrims have to be state-sanctioned. I called my relative from a phone booth in the Jaigahan bazaar.

After our greetings, I asked, "Are you going?"

"It is not so simple being a Muslim in India," my cousin told me.

His wife had to get a new passport before she could apply for a hajj visa. "They see a Muslim name, and they delay the application," my cousin said.

It continued like that for days. Finally, I had to admit to myself that I was not going to be able to do the tricks in this circus—at least not that year.

THE WALL OF WAHHABISM

KARACHI, PAKISTAN—Later that year, I found myself in the most unlikely of places. I was living in a posh neighborhood in Karachi, Pakistan.

When Muslim hijackers killed thousands in the September 11 terrorist attacks on the World Trade Center and the Pentagon, I went to Pakistan as a journalist for *Salon* magazine. I had been writing my book in Morgantown but put it aside. I was a Muslim of the West. Somehow, I thought that through journalism I could bridge some of the critical failings that had led to the violence of September 11. I had last been in Pakistan under very different circumstances. When I was twenty-seven, I met a Pakistani man, projecting onto him my deep desire to reconcile the dissonance in my life between East and West. Like me, he had run high school cross-country, and he lived a bifurcated life: born in Pakistan, he was raised in Paris and settled in Washington, D.C. Over the years my mother had warned me: "If you marry an American, your father will have a heart attack." Muslim guilt set in. Within weeks I left an American Lutheran boyfriend who loved me fully, said he was willing to convert to Islam, and was ready to learn Urdu. I got engaged, sold the condo I had bought off Chicago's Lake Shore Drive, and moved to Washington to prepare for a wedding in Islamabad, Pakistan. The deeper voices of my religion were speaking to me: the ban on Muslim women marrying non-Muslim men, the disapproval toward sex before marriage. I was looking for a reunion between my two selves.

Within weeks after the wedding, however, I knew I hadn't made a suitable match for me. I fell into depression, and my husband left the marriage, withdrawing from our joint bank account the proceeds from the sale of my condominium. After that disillusioning experience, my pendulum swung to the West, and I escaped into beach volleyball, my work, and casual dating. Still, I couldn't escape my programming. I threw hopes of matrimonial bliss on every man I dated. On my trip to Pakistan after September 11, I landed in Karachi alone, on a grim search for the family of a suicide victim. I was reporting a story on how despair over dismal economic opportunities was driving record numbers of men to commit suicide in this Islamic country where religious clerics ruled that suicide was a one-way ticket to hell. I didn't think their rulings allowed compassion for the real chemical imbalances that actually drove people to commit suicide. To me, their rule was as compassionate as condemning a heart attack victim to hell. The helplessness and hopelessness that led Muslim men to acts of violence against themselves seemed to have much in common with the imbalance and frustration that was leading other Muslim men to commit suicidal acts of terrorism, such as the 9/11 disaster.

To my surprise, in the midst of my reporting I fell in love with a young Pakistani man. We met one Sunday afternoon when I went to visit the son of a local newspaper columnist, a family acquaintance, at a weekend getaway called French Beach. We started seeing each other, as openly as if we were dating in the West. The candor surprised even me, but I got caught up in the emotion. He told a friend, yelling into his cell phone as we entered an elevator at the Karachi Sheraton Hotel and Towers, "I've met the woman I'm going to marry!" His proclamation might have seemed frivolous to some, but I took it seriously. I had been traveling the Indian subcontinent for two years searching for divine love. Just when I had given up, I seemed to have found it.

During the first days of January 2002, I rented a villa with jasmine flowers in the garden to write my book and pursue my romance. Again, I saw the two aspects of myself in this Muslim man whom I had met. Even in love, I was seeking truth and wisdom.

I was also again trying to do my hajj. This time I had a potential assignment from *Outside*, an outdoor adventure travel magazine, to experience the hajj in the most adventurous way that I could. For years as a *Wall Street Journal* reporter, I had thought I could do an unconventional profile on the big business of the pilgrimage. It amazed me that you could do the hajj and also get Hilton honor points. The magazine assignment seemed a good way to report this side of the hajj phenomenon. I was struggling with the question of whether to go as a journalist or as an ordinary pilgrim. To go as a journalist, I would have to get a journalist's visa from the Saudi Ministry of Information. I was fortunate to know a fellow *Wall Street Journal* reporter who had penetrated Saudi Arabia's bureaucracy and reported from there. His name was Daniel Pearl, and we'd been friends since the summer of 1993, when I was reeling from my failed three-month marriage to the man from Pakistan. Close in age, we bonded immediately and built a close friendship over office pranks, beach volleyball, and immigrant parents who wanted us to marry within our ethnic and religious roots. Danny made a decision to choose love over religion and ethnicity; he married an exuberant French Buddhist woman, Mariane Vanneyenhoff, born to a Cuban mother and Dutch father. Danny became the South Asia bureau chief for the *Journal*, and he and his wife flew straight to Pakistan after the planes hit the World Trade Center. I called him in Islamabad, where they were staying.

"Should I go as a journalist or an ordinary pilgrim?" I asked him.

Danny told me a Saudi information chief was a man named Prince Turki bin Sultan bin Abdul Aziz.

"He's a real turkey," Danny chortled, in a joke he admitted was obvious. More seriously he said, "Stay away from him."

These weren't ordinary times anymore. On the ground as journalists, we were trying to avoid the dangerous nexus between political power and world events and not get entrapped by nefarious personalities or hidden agendas. Another Prince Turki—Prince Turki bin Faisal bin Abdul Aziz al Saud, the son of King Faisal, ruler of Saudi Arabia from 1964 to 1975—had been making the news under less than stellar circumstances. From 1977 through August 2001, he had been head of Saudi Arabia's Department of General Intelligence. Prince Turki was considered a potential heir to King Fahd, the ruler of Saudi Arabia since 1982. He abruptly resigned late in August 2001 and was replaced by an uncle. Journalists had raised questions about his being affiliated with the puritanical and anti-American wing of the royal family sympathetic to Osama bin Laden. The U.S. independent commission reviewing the September 11, 2001, attacks on America later reported that the Clinton administration had turned to Prince Turki bin Faisal for help in getting bin Laden expelled from Afghanistan after the Saudi government successfully thwarted a bin Laden–backed effort in the spring of 1989 to launch attacks on the U.S. military stationed in Saudi Arabia. The report said that, as Saudi intelligence chief, Prince Turki bin Faisal, using "a mixture of possible bribes and threats," received a commitment from Taliban leader Mullah Omar to hand over bin Laden. But Omar broke that pledge during a September 1998 meeting with Prince Turki and Pakistan's intelligence chief. "When Turki angrily confronted him, Omar lost his temper and denounced the Saudi government. The Saudis and Pakistanis walked out," the report would say.

From Danny's warning alone, I decided to proceed as an ordinary Muslim pilgrim. I soon learned that that wasn't going to get me anywhere fast.

Outside editor Stephanie Pearson connected me to a Saudi guide she found through a documentary producer and convert to Islam, Michael Wolfe, also the author of a book about the hajj. I had never met Michael, but I would one day feel blessed that I got to know him on this journey. The guide was a Saudi architect, Dr. Sami Angawi, who traced his lineage to the prophet of Islam. With a PhD in Islamic architecture from the School of Oriental and African Studies at the University of London, he lived in a city called Jeddah, the main arrival point for pilgrims. He founded an organization called the Amar Center for Architectural Heritage to preserve the traditional history of Saudi Arabia, and he was a former adviser to the government on the hajj. Stephanie envisioned me doing my hajj in the path of the prophet Muhammad, camping out and trekking through the hajj, just as a caravan might have done at the time of the prophet. Sami seemed to be the perfect person to guide me in re-creating the prophet's steps. But I realized we weren't planning a jaunt through Yellowstone Park when I started asking Sami questions that would help us realize the vision.

"Can we camp out?" I asked him.

Sami answered with a question: "How old are you?"

That seemed like an odd response, but I answered anyway. "Thirty-six."

"And your father?"

"He's sixty-eight."

"I don't know. It might be suspicious. They might wonder what a man is doing with a woman in the desert."

Okay. I was speechless. My father was not only old enough to be my father; he *was* my father. Such considerations were beyond my cultural sensibilities.

I envisioned following the path of the prophet Muhammad's pilgrimage on camelback. "That will be difficult," Sami warned me.

"Why?"

"The Saudis have destroyed so much in the name of Wahhabism."

"*Wahhabism?*"

Before Wahhabism became a household term, I didn't know what this word meant. I hadn't grown up with a sense of the differences between Muslims. I always thought a Muslim was a Muslim was a Muslim. Sami explained the history to me. Wahhabism is the ideology of Saudi Arabia. What I was going to learn was that just as Christianity has different denominations, Islam is not a monolithic religion. It has two sects, the majority Sunni and the minority Shi'a. The Sunni history was marked by scores of schools of jurisprudence called *madhabs.* Four schools had survived into the modern day: Hanafi, Malaki, Shafi'i, and Hanbali, named after the scholars who led them. Mystical Islam, or Sufism, evolved both in peace and in conflict with these schools of jurisprudence. Wahhabism sprang out of the Hanbali madhab, and it is more puritanical and rigid than most other schools of Islam practiced around the world. It arose in the eighteenth century through a religious reformer, Muhammad Ibn 'Abd Al Wahhab, and later the ruling Saud family had made it the law of the land in order to use religion to control the masses. Following this brand of Islam, Saudi Arabia doesn't allow cinemas or theaters because it considers most entertainment frivolous and often *haram,* or "unlawful," according to Islamic standards. The country not only forbids churches and synagogues but prosecutes Christians for holding religious services in their homes. To Sami's professional horror, Wahhabi clerics had dismantled the prophet's house in Mecca to clamp down on anything that might be interpreted as worship of the prophet Muhammad. (In Islam the prophet Muhammad is considered a very great man, but still a man; in the Wahhabi interpretation, there should be no tokens of his life that Muslims could worship.) In the process, Wahhabi clerics had destroyed other relics from the lifetime of the prophet Muhammad. Their religiosity is akin to the Puritan practice of Christianity in early American history. When I next called home to West Virginia, I shared with my father what I'd learned from Sami. He lamented the loss of so much history in the name of religion.

Trying to get a visa, I realized that I was going to have to overcome other expressions of Wahhabism before even landing in the desert with my father. An official at the Saudi consulate in Karachi told me my passport had to be submitted with the passport of my mahram. The process seemed ridiculous. After all, I had traveled freely through Pakistan, a Muslim country, without a mahram. Just three months earlier, I had applied for a visa into the Taliban's Afghanistan without a mahram. (I was turned down, but for other reasons, not least of which was the looming war with the United States.) My travel experience was just more evidence that as much as the puritan Wahhabis insisted their path was the *only* true Islam, there were alternative ways of practicing Islam.

As a last-ditch effort, I visited the consulate of Saudi Arabia. A polite Pakistani man behind the counter ushered me into a meeting with a hajj official. Sitting in his office, I eyed the piles of hajj applications around me.

"Isn't there a way that I can get my visa through the Karachi consulate and my father get his visa through the Washington, D.C., embassy? We could meet first outside Saudi Arabia in, say, Amman, Jordan, and travel into Saudi Arabia together?"

"I am sorry. That is not possible." He explained the rules: we had to apply together through a registered tour operator, and the tour operator would submit the application to the government of Saudi Arabia.

Trying to work every angle, I asked, "Okay, *you* can't give me permission. But how might I get permission?"

He looked at me curiously, as if wondering why his denial wasn't good enough. He scribbled on a piece of paper and passed it to me. "Hajj Ministry." This was the government office in charge of Pakistani pilgrims. I could try to get a tour operator to agree to let me go with his group. Every time I called a local travel agency, however, I got the run-around. Meanwhile, in Morgantown my father was so stressed about arranging the logistics of traveling with me in Saudi Arabia that he was driving my mother crazy.

BETRAYAL AND A TURNING POINT

KARACHI—As the start of the hajj crept closer, I was still desperately trying to pull off the trip when Danny and Mariane came to visit. Mariane was five months pregnant, and they had just learned the day before that their baby was a son. That night we exchanged war stories, quite literally, listened to Tom Petty, Bruce Springsteen, and a Pakistani Sufi rock band, Junoon, and tried unsuccessfully to watch DVDs on Danny's laptop.

The next day, January 23, 2002, Danny left for an interview he thought he was going to have with a spiritual leader named Sheikh Gilani. He was investigating reports that Sheikh Gilani had ties to Richard Reid, "the shoe bomber" who had tried to blow up a trans-Atlantic flight from Paris to Miami by lighting a fuse connected to explosives hidden in his sneakers. Unbeknownst to Danny, a Muslim militant by the name of Omar Sheikh had hatched a plot to kidnap him. In the early evening, Danny stopped answering his cell phone. Worried, Mariane and I began a mission that night, searching for Danny. The next morning, after we alerted the *Wall Street Journal*, the U.S. government, and Pakistani police, my house became a command center for the investigation. My boyfriend had been with us the first night, but when he came over the second night after Danny's disappearance he told me that Pakistani intelligence officials had visited him to find out what he knew about Danny and me. The visit frightened him as well as his parents and friends. He wouldn't come around anymore. I wept that night, privately, in a walk-in closet where I knew nobody would find me. And then I wiped away my tears and focused on only one goal: finding my friend. With Mariane, I crossed physical boundaries rarely breached in that culture. We lived alone in that space that women rarely claimed without a chaperone—their homes—and we worked alongside FBI agents and Pakistani antiterrorism specialists trying to piece together the clues left by Danny's kidnappers.

As the first week of our search ended, I wrote to my father to assure my family that I was safe, with armed police escorting us everywhere.

"We will find Danny," I wrote. "Every day we are one step closer."

As Mariane and I entered the third week of our desperate search for Danny, I realized something shocking: I might be pregnant. A pregnancy test confirmed my suspicion. I was shocked. I had never gotten pregnant before. I didn't know what to do. I didn't wear a wedding ring, but I didn't feel as if I had done something wrong. I had loved my boyfriend deeply and surrendered myself to him. Even if my assumptions had been wrong, I loved him when I made this baby. He had abandoned me, but that was not because of my failure. It was because of his fears. I called my boyfriend and asked him to visit me. He arrived that night, and I took him to my bedroom. "I am carrying your baby," I told him, sitting on the edge of my bed.

He looked at me stunned. In a pause that I filled with so many dreams, he sucked his breath in hard and said, "I have to go."

The truth revealed itself. He didn't want me to keep the baby, and all of his fanciful talk about marrying me disappeared. Despite my intellectual confidence in myself, I felt completely illegitimate. Within me was an American woman who believed in free will and thus knew that I had the right to keep my baby and raise him with my head held high. But the voices of my religion's traditions also spoke strongly inside of me. I was consumed by the shame of ignoring the rulings of sharia, the "divine Islamic law." For reporting I had done on the subculture of sex, drugs, and nightclubs in Pakistan, a leader of Jamaat-i-Islami, one of Pakistan's religious parties, told me that the laws policed even sexual intimacy between consenting adults. In 1979, he told me, Pakistan passed laws based on *hudud,* or "boundaries" for moral conduct.

I knew these rules were used to control us, but now I learned that violating them could have more serious consequences. My situation could land me in prison.

To me, these laws emblemized a deeper crisis of self-determination for women in Islam. Women in Islam are so very much defined by hudud. These hudud are used to control everything about our lives, from our sexuality to where we can pray in the mosques that are our places of worship. By other names, these types of boundaries have also defined women throughout time in other cultures and religions, including Judaism and Christianity. So often religion is used to impose boundaries that ultimately deny women rights that have now been articulated in the Universal Declaration of Human Rights: the right to self-determination, the right over our bodies, the right to travel freely. These religiously imposed boundaries directly affect a woman's economic life, identity, sexuality, and political power. But when I discovered I was pregnant, I realized that the

deepest boundaries we have are within ourselves. We are often most constrained by the fears that keep us from crossing the boundaries.

Tragically, in the fifth week of our search, we found out that Danny's cap-
tors had slaughtered him. They had taken a knife to his throat and decap-
itated him. They titled a videotape of his murder *The Slaughter of the
Spy-Journalist, the Jew Daniel Pearl*. It lasted three minutes and thirty-six
seconds. I was in disbelief.

The next day the phone in my house rang. It was Mohammedmian
Soomro, the governor of Sindh, the province in which Karachi lies, and
his wife, Khadijah Soomro, expressing condolences. They were on the
hajj that I had been trying to join. Mariane and I had met them in our effort to lobby for political muscle behind the search for Danny. They had
been kind and generous, and Khadijah Soomro had given us shawls. In
Mecca they were celebrating a holiday called Eid ul Adha, which marks a
pivotal moment in the life of Abraham. Like Jewish and Christian children, I was taught that God told Abraham to slaughter his son as a sign of
his devotion to God. Modern-day Muslims slaughter goats, sheep, and
even camels as a symbolic gesture of their willingness to sacrifice in devotion to God. Outside my house in Karachi, goats had been bleating for
days, tied up for the sacrifice. In the hours since we received the news of
Danny's murder, their noises had grown increasingly dimmer.

The governor and his wife were somewhere in Saudi Arabia. I had no
idea where. I still didn't understand the confusing litany of rituals that
filled the hajj. I wondered about Islam in the world. In the name of religion, men punished—and even killed—women like me who were unmarried and pregnant. In the name of my religion, men hijacked planes and
flew them into the World Trade Center, murdering thousands of inno-
cents and changing the course of histories, both personal and global. In
the name of my religion, men had slaughtered Danny, a young man with
dreams no more complicated than to buy a double bass for his fortieth
birthday, love his wife, parents, and sisters, and nurture his son.

Instead of Danny's dreams being realized, police were interrogating four
young men who were charged with plotting Danny's kidnapping. They
considered themselves devout Muslims. While planning Danny's kidnapping, they had interrupted their strategy sessions to bow their heads toward Mecca for the obligatory five-times-a-day prayers. His murderers
videotaped Danny talking about his Jewish heritage, which, in their puritanical Muslim hatred of Jews, was enough to sentence him to death.

Among Danny's last words were: "I am a Jew." To my shock, Danny's murderers slandered the name of Islam by killing in its name.

On the dawn after the news of Danny's murder, I was standing in the abyss of a darkness wrought by man's distortion of religion. I was engulfed in a pain that made me feel the angels crying when it rained that morning. I was angry. I was afraid. I was sick to my stomach when I even dared allow myself to feel. Yet, in that sacred space of my womb, life had been created.

In my computer "in-box," I had received an e-mail with rental prices for campers in Saudi Arabia. But early on in the desperate hunt to find Danny, I had told the editor at *Outside* that I was putting aside my hajj planning. She understood immediately. By then, my destiny was clear. It wasn't yet time for me to do my pilgrimage. Would there ever be a time? Michael Wolfe, the author and producer who had introduced me to Dr. Sami Angawi, had a philosophy that answered my question. "Hajj is in the heart. You don't go until you're supposed to go."

In my heart, I felt fear and loathing for my religion. Could I remain in a religion from which so many people sprang spewing hate? Could I find space in my religion for my kind of woman? Could I remain a Muslim?

I didn't know the answers, and I wondered if it would ever be time for me to venture into the heart of Islam.

BIRTH AND REBIRTH

MORGANTOWN, WEST VIRGINIA—It took the great gift of a child in utero to put me in the right place in my life to do my pilgrimage. It seemed that the answers to my questions about my religion lay within me. Answers to all of the great mysteries are to be found in our inner selves, but in my case this was literally true.

Throughout my pregnancy I tried to make peace with my boyfriend so that we could marry and make the nuclear family that so many people in my religion and culture expected. But my baby's father broke promise after promise, leaving me empty and depressed. I spent most of my pregnancy in Paris, helping to bring Danny and Mariane's baby, Adam, into the world. I had a certain amount of responsibility and guilt to handle.

"As a Muslim, I want to welcome Danny's son into the world," I told a man Mariane and I called "Captain." He was an honorable Muslim law enforcement officer who led an antiterrorism team in Karachi. Even though the hajj had bypassed me one more year, I was still trying to get glimpses into this mysterious experience. Captain told me that his faith had been renewed when he did the hajj with his wife years earlier. Now, to defend his religion, he fought to stop men who killed in the name of Islam; he wanted to represent Islam well in the world.

I knew of what he spoke. I didn't pretend to be a model Muslim according to Islamic standards for rituals and external appearances. I didn't pray the requisite five prayers a day. I didn't cover my hair. And, yes, with my baby as evidence, I had sex outside of marriage. Although I had a firm faith in a divine force, I didn't invoke the name of a God who judges, punishes, and rewards. I tried simply to live as a good Muslim with humanitarian values, in the same spirit as a good Christian, Jew, Hindu, or Buddhist. I didn't lie. I didn't cheat. I tried not to hurt others. I tried to live sincerely. And I wouldn't think of taking someone else's life. Thus, I was reeling from the shock of the darkness in the world when I decided to return to my hometown at the start of my ninth month of pregnancy. I was still trying to understand how some men could take the lives of others, and how another man could want me to take the life that sprang up within me, all in the name of preserving their image of propriety.

Morgantown was the only refuge I could envision for myself in the world. It was a place where I felt I had roots. Since my family moved there when I was ten, my parents had never moved away. After I left the nest, first going to Washington, D.C., then San Francisco, Chicago, New York, India, and Pakistan, I always returned to Morgantown. I had been running the same route around my neighborhood since I was twelve, turning right off Riddle Avenue to avoid a mammoth hill that was the hallmark of the landscape of our little town. Morgantown was where I learned to be a writer and honed my skills at the local newspaper, the *Dominion Post*. In case I had any doubts, the city had made the top ranks of best small American cities in which to live.

However I intellectualized about what I had experienced, I was also devastated, broken, and raw when I returned to Morgantown. When I arrived, my mother and father hugged me and my swollen belly without a word of reprimand. "We love you. We love our grandson," they said. I sank into my parents' loving embrace and settled into a room my father had added to their new house. They cared for me and tried to return me to physical and mental health so that I could be strong for my son.

My son entered my arms on October 16, 2002, on the sixth floor of Ruby Memorial Hospital, five days before the nine-month anniversary of

Danny's disappearance. When I gazed at my son, I knew divine love, I knew heaven, and I knew God. I had been blessed with life springing forth from the midst of death. Because of my parents' love, I was able to love my baby with a full and open heart. They were expressing the principle that Islam had taught them best: love. I looked at my son, and I knew that the name I had chosen for him was perfect: Shibli Daneel Nomani.

Shibli Nomani was the revered name of a paternal ancestor who had been a reformer and scholar in the Indian Muslim world at the turn of the twentieth century. As I was told, Shibli means "lion cub" in Arabic. I chose Daneel for my son's middle name because I wanted to pay homage to the spirit of my friend Danny. Over the nine months of my pregnancy I had come to firmly believe in the meaning of the name Daniel: "God is the judge." It seemed that, in the rush for moral supremacy in our world, so many have forgotten this simple message. Danny's murder and the rejection of my pregnancy by puritanical Muslim standards of decency told me that the world is in need of more pluralism and tolerance. It was for this reason also that I gave Shibli a version of the Jewish name Daniel. Shibli Daneel—Daniel and the lion cub—captures an important message of the Bible story about Daniel in the lions' den. A king sentences Daniel to death in the lions' den because Daniel refuses to worship him. Daniel's spirit transforms the lions into friends, and the message of the story is clear: God, not man, is the judge of what transpires in our hearts.

According to Islamic tradition, Shibli's father was supposed to whisper the call to prayer, or *azan*, into his right ear. It is the same call to prayer that springs forth from the minarets of mosques throughout the Muslim world. It is heard five times a day in Mecca. But Shibli didn't hear this proclamation from his father, who had broken his promise to be with me during delivery. Instead, my father performed the sacred duty of uttering the azan into Shibli's ear on the sixth floor of Ruby Memorial Hospital.

Allahu Akbar (God is great)
Allahu Akbar (God is great)
Allahu Akbar (God is great)
Allahu Akbar (God is great)
*Ash-hadu alla ilaha illa-llah (I witness that there is none worthy of worship
    but Allah)*
*Ash-hadu alla ilaha illa-llah (I witness that there is none worthy of worship
    but Allah)*

Ash-hadu anna Muhammadar-Rasulullah (I witness that Muhammad is the
    messenger of Allah)
Ash-hadu anna Muhammadar-Rasulullah (I witness that Muhammad is the
    messenger of Allah)
Hayya 'ala-s-salah (Come to prayer)
Hayya 'ala-s-salah (Come to prayer)
Hayya 'ala-l-falah (Come to success)
Hayya 'ala-l-falah (Come to success)
Allahu Akbar (God is great)
Allahu Akbar (God is great)
La ilaha illa-llah (There is none worthy of worship but Allah)

"Come to prayer. Come to prayer. Come to success. Come to success," I said to myself. The words echoed in my mind over the next
weeks as I tried to reconstruct my life. I was drawn again to do the hajj.
To complete the hajj is to emerge "reborn," in the tradition of the great
evangelical Christian revivals. What more could I seek? I felt that I had
made many mistakes that I would not reverse or erase even if I could,
because they had all led me to the creation of my son and they were
necessary for me to understand what I truly believed. I had to acknowledge that they also brought me great pain and even sorrow. According
to the rulebook of Islam and most of the other religions, I had sinned. I
had broken the moral code of my religion. The father of my child would
not marry me because I was not acceptable in his family's traditional
ethos. The hajj is supposed to be for the pious. But I knew I needed a
new beginning. Living with my parents and my all-important mahram,
the logistics of doing the hajj were now possible. I made my decision. I
would go on the hajj.

And, sleep deprived and perpetually exhausted, I realized that I was
going to raise my son alone. I called my boyfriend in Pakistan one night
when Shibli was just weeks old. I misdialed the country code and city
code and dialed the 911 operator. The West Virginia State Police dispatcher answered the phone. I hung up. The phone rang. It was the dispatcher calling back to make sure I didn't have an emergency. "Is
someone holding a gun to your head?" she asked. The epiphany hit me.
Someone was holding a gun to my head, and she was I. I had to let go and
start a new life with my son. I had always thought of a pilgrimage to
Mecca as an assignment. But I realized that maybe I was ready for something deeper. I had started reading about a woman named Hajar. She was
one of the few hits I got when I did a Google search of "Islam and single

mother." I didn't know her story, but somehow it was intertwined with the hajj, and I knew I needed to know more.

Now the question was whether I had satisfied the spiritual prerequisites for doing the hajj. "Am I worthy?" I wondered, curling my body around my sleeping baby, caressing the top of his head with my cheek. His newborn warmth soaked straight through to my heart.

FEAR AND DOUBTS

MORGANTOWN—I started filling out visa applications for the hajj as Shibli's first teeth started emerging in his third month. The warning of a death sentence screamed at me from the visa application for Saudi Arabia. It was a proclamation—loud and clear—that anyone caught with drugs faced the death penalty. I was clean on that account, but the message underscored the reality of a society that strictly punishes lawbreakers.

I would be a lawbreaker. Even more than Pakistan, Saudi Arabia punishes women like me who have children outside of marriage. These countries' strict interpretation of sharia says that unmarried men and women who have sex are to be whipped and married men and women found guilty of having sex outside of wedlock are to be stoned. The crime: *zina*, the Arabic concept of illegal sex. If Nathaniel Hawthorne's scarlet letter was an "A," would I have to wear a scarlet "Z"?

I knew my pilgrimage would be a much more cautious journey than I had planned the year before because, in my heart, I knew I had to go with Shibli. Some of my friends warned me that he was too young to travel, that his immunities were still developing. But I knew that I wanted him with me because he was the result of the best in me. He lived because I had chosen life over fear. He smiled because I had chosen happiness over shame. He grew because I believed the present and the future define us, not the past. He was the result of my efforts to be a better person, to flow toward the divine.

The hajj represented a danger to this new being because it is a physically arduous journey marked by fires and stampedes. In 1975 there was a fire in a tent colony outside Mecca that killed thousands. In 1987 the Saudi government gunned down about 400 unarmed Iranian pilgrims protesting its rule. In 1990, 1,426 pilgrims were crushed to death in a stampede in a pedestrian tunnel leading from Mecca to Arafat. In 1994, 271 pilgrims were trampled in a stampede. In 1997, 343 pilgrims burned to death and another 1,500 were injured in a blaze that roared through 70,000 tents outside Mecca. The air was left thick with the smell of smoke, and burned-out buses, charred water bottles, and other blackened debris littered the ground. In 1998, 119 pilgrims were crushed to death in a stampede. In all of these incidents, spokesmen for the government blamed divine predetermination.

To die in Mecca during the hajj is considered by many to be a blessing, and people will sometimes abandon personal safety for faith, creating dangerous situations for those like myself who aren't particularly interested in dying. I knew that I didn't want to lose my son to human caprice after I had overcome so much to bring him into the world. I had to carefully safeguard the precious health of my baby, but I also knew I had to take the risk.

One of the main purposes of hajj is to submit completely to God. It is like taking an oath of citizenship. For Muslims, our Ellis Island is Mecca. Our Statue of Liberty is the Ka'bah, a peculiar center of devotion by any stretch of the imagination. Its name means "cube," appropriate for its box shape. The Ka'bah, which sits in the middle of a courtyard surrounded by a marble palace to God, is quite unremarkable architecturally. It's a big square building about the size of half a football field. The Ka'bah reminded me of the poem written by Emma Lazarus in 1883 and since memorized by so many American schoolchildren, such as my niece Safiyyah when she was a second-grader at North Elementary School in Morgantown.

> *Give me your tired, your poor,*
> *Your huddled masses yearning to breathe free.*
> *The wretched refuse of your teeming shore.*
> *Send these, the homeless, tempest-tost to me.*
> *I lift my lamp beside the golden door!*

Symbolizing heaven and earth and everything in between, the Ka'bah beckons the spiritual pilgrims of Islam. It is supposedly the house of God on this earth. Wherever we are on earth, all Muslims are supposed to pray five times a day with their prayer rugs laid out toward the Ka'bah in Mecca. To pray in front of it on the hajj is supposed to reinforce a Muslim's belief. When we return to our homes, we have new status in life and are called by a new honorific: men add *hajji* to their names, and women add *hajjah*. In many parts of the world relatives and friends greet returning pilgrims with garlands of roses and jasmine. The pilgrimage symbolizes a

transformation. As I thought about going, I had to admit my own limitations: I didn't think I would be so deeply touched by this religious exercise
that I would emerge transformed.

DEPARTURE AND THE FAITH OF MY PARENTS

MORGANTOWN—Unlike me, my father was thrilled that Saudi law forbade me to travel alone to Mecca and required that I travel with a
mahram. This meant he got to do the hajj. For my father, doing the hajj
was the culmination of a life committed to Islam.

Born on June 14, 1935, in India, my father grew up in a traditional
Muslim family with a strong matriarchal figure, his mother, Zubaida Nomani. Slight in figure, he is large in spirit, ready to drive cross-country at a
moment's notice if I asked, my friends would joke. He talks with conviction until I, testing my limits, challenge him so much that he retreats to
try harder. He is so prone to passion and dramatic flair when he talks that
somebody once went to break up what he thought was a fight between my
father and a friend. His friend joked, "No, he's just part Italian." His intense gaze reveals his convictions and integrity. When he set off for his
first job as a lecturer at Osmania University in Hyderabad, India, he laid
his prayer rug toward Mecca and said a prayer to bless his work. Throughout his lifetime he would turn to Mecca for peace, salvation, and prayer.

He faithfully adhered to the five pillars that Islam prescribes for Muslims. The first is the proclamation of the *shahada*, or the "declaration of
faith" that there is one God, with Muhammad as a prophet. My father was
firm in this belief. The second pillar is the five daily prayers. My father did
his prayers with devotion but wasn't fanatical about it. If he missed a
prayer, he made it up later. The third pillar is *zakat*, or charity to the poor.
Each year, my father calculated our household zakat obligation, not unlike
figuring out a tax bracket, and signed a check over to the Morgantown
mosque or to one of his sisters for charity work in Pakistan or India. The
fourth pillar is fasting from sunup to sundown during the holy month of
Ramadan, which shifts on the Western calendar year because its dates are
based on a lunar cycle, with about 354 days in a year. As a professor of nutrition, my father had taken the commitment to fasting to heart by becoming an expert on nutritional issues related to Ramadan fasting. Before
blogging ever became a part of modern society, my father wrote his own
HTML script to create the *International Journal of Ramadan Fasting*.

The fifth pillar is hajj. As I'd discovered in Pakistan, Saudi law requires that pilgrims arrive with a Saudi-sanctioned tour group to fulfill the fifth pillar. My father searched the Internet and scrolled through all of the options. Trip Advisor listed "Mecca Hotels" and "Things to Do in Mecca": of two items, the Holy Mosque got top billing. (The second was the cave where the prophet Muhammad got his revelations.) First, we had to find a tour group. Who, we wondered, would give us the balance we sought of spirituality and scholarship? My father came to me with an answer: the Islamic Society of North America. I only vaguely knew about this group. They were sort of like the NAACP of American Muslims, but they seemed to represent the immigrant generation more than my generation. The Islamic Society advertised a hajj tour on its website, and its list of guides included scholars. I liked that.

As we were considering joining its tour group, my father excitedly called me to the TV. C-SPAN was airing footage from the society's recent national convention. I studied the footage. It seemed that all the women were wearing head scarves. That signaled to me that this was a more conservative crowd that believed women *had* to cover their hair. I didn't see women who chose not to cover their hair like my mother and me. I wondered whether head scarves were required, or whether the society passed them out at the door to any women who arrived with their hair bare. Either way, the lack of diversity made me feel uncomfortable. Still, the society's tour group seemed like the smartest one we could join. With some trepidation, I gave my father the green light to sign us up.

My mother decided she would accompany us to help me with Shibli. My mother is petite and striking, with strong cheekbones and proud eyes like a Cherokee Indian's. She cuts through ambiguity with precise analysis, sometimes to my annoyance when I'd rather waste time in nuances. She moves slowly, with poise and dignity, from the morning, when she sips her Lipton tea, to the night, when she studies Fareed Zakaria columns in *Newsweek*. Beneath her calm is a fiery spirit and independent mind that I inherited. She had had a wider experience with religion than my father. As a child, she was taught by Catholic nuns at a convent called St. Joseph's in a lush hill station in India. The puritans in Islam considered such an education corrupting. My mother considered it expansive (as well as irritating when the nuns got angry). She grew up in a conservative family that devoutly practiced the five daily prayers, the full month of fasting, regular alms giving, faith in the monotheistic creed, and pilgrimage. Although she had been a follower of Islam all her life, dogmatic adherence to faith and rituals turned her off. She had broken many traditions to become an entrepreneur, running a popular boutique in downtown Morgantown.

In contrast to my father, my mother didn't particularly care about doing the pilgrimage. My father wept for joy at the thought of going to Mecca. My mother rolled her eyes. Although the pilgrimage is important to many, she saw solving the inequities and the injustices in the Muslim world as the real priority facing Muslims. For my own part, I wasn't sure about the hajj, but I wanted to experience it to decide for myself. For her, Islam was a private act of faith. She had never been a part of my father's community building. She believed in the shahada, the proclamation of faith, but she didn't consider Muhammad a greater prophet than Jesus or Moses, just the last in a line of men with monotheistic messages. This is what the Qur'an itself even teaches about the prophet, but distortions of Islam's teachings have established a religious pecking order, as happens with the fundamentalist expressions of all religions. Having grown up with prayer enforced strictly, my mother rarely unfurled the prayer rug, to the disappointment of my father. To her, prayer doesn't know a time clock or a prostration. Life is prayer. Moreover, zakat was always a source of conflict between my father and mother. She didn't believe my father had to send his money overseas when there were plenty of poor people in America.

"Are you excited about going on the hajj?" I asked her one night as she stood at the kitchen sink, where she seemed to spend so much of her life.

"Not really," she said bluntly, turning a bottle of dishwashing liquid upside down on a sponge.

I fell somewhere about halfway along my parents' spiritual spectrum. I appreciated my mother's honesty. It was a refreshing alternative to blind faith and the absence of critical thinking. I had to admit, however, that her response disappointed me. I wished she were more enthusiastic. But I imagined she knew what was in store for her.

My mother was the buttress to the castle that was our home. My brother had married, and he and his wife and their two children lived with my parents in an extended family environment my mother kept running. After I had my son and returned home from the hospital with him, I was bedridden from the physical trauma of the C-section. Without my mother, who would lift my son from her bed in the room adjoining mine and carry him to me, I don't know how I would have been able to nurse him. Because my brother's wife was studying around the clock in her last semester of nursing school, my mother worked double time at her boutique during the day and then after 5:00 P.M. started her second shift, keeping the house in order, getting dinner on the table, and tutoring my brother and sister-in-law's eleven-year-old daughter and nine-year-old son so they could excel in school. That night as we talked she turned from the dishes to dinner.

My father joined us and spoke about the harassment and persecution the prophet Muhammad faced in Mecca when he started preaching Islam.

"Oh! The prophet," he sighed. "How they all suffered!"

"Oh, Dad!" I exclaimed, irritated and embarrassed at his emotionalism. At that time, I had little clue about Islam's history. I didn't even know in which century Islam began (the seventh).

My mother watched my father and shook her head without sympathy for his tears. "That drives me crazy," she said. She couldn't conceive of feeling anything close to my father's devotion to historical icons. Religion had largely alienated my mother because she had seen it used to justify acts that were devoid of kindness and human compassion. My mother looked over at Shibli, curled in my arms. To her, it would be a far greater act of devotion to care for my newborn son than to circumambulate a cube-shaped building. "I'll stay in the hotel room and do the rituals another time," she declared.

I rolled my eyes and sighed, "Oh, Mom!" My mother had the notion that fulfilling the ritual would be dangerous and stressful. I disagreed, but she was more sensible. My mother also had more knowledge than I did of the extent to which people forsake good judgment in pursuit of religion and ritual. Growing up with a deep expression of Islam in her family gave her a broad understanding of Islam, but she was alienated by the cultlike power of religion. My father had gone to Mecca once in 1967 for the umrah, the pilgrimage done outside the time of the hajj. (It doesn't count as the required pilgrimage but does bring extra blessings.) When he returned, he wrote to my mother, who was in India with my brother and me, saying that he wanted to embrace religion more. But my mother rejected what she saw as the intoxication of religious devotion, to which the pilgrimage had made my father more susceptible.

Because my sister-in-law was going to be in school full-time, she suggested that my niece and nephew join us on the hajj. It thrilled me that they would join us, balancing our adult sensibilities and making our team truly intergenerational. Eleven years old, Safiyyah was in the sixth grade and wore Mudd jeans, danced to Queen Latifah, and patted her eyelids with eye shadow. Her brother, Samir, was nine and in the fourth grade. He hung a portrait of Buddha on the wall not far from an image of Nintendo's

Super Mario and rap star 50 Cent. Like Shibli, Safiyyah and Samir spent their early years at my mother's boutique on High Street, sitting in her armoires and playing hide-and-seek among hand-embroidered silk skirts.

The logistics were falling into place, but apprehension plagued me. I was a modern Muslim woman and new single mother, making the journey with my elderly father as my male escort; my mother, a successful business-woman who was shuttering her boutique for the first time in two decades to make the journey; my newborn son, still suckling at my breast; and my hip niece and nephew. I was a postmodern woman in a religious culture with many premodern dispositions. Could I find a place for myself within my religion? My parents, in their different ways, had faith that I could.

PART TWO

STARTING THE PILGRIMAGE

February 2003

AN INTRODUCTION TO COMMUNITY

JEDDAH, SAUDI ARABIA—As we stepped into the visitors' lounge at the Jeddah airport, a sea of poor, dark-skinned men and women from Sudan floated before me, a jarring vision in the first moment that I stepped into Saudi Arabia. We were in a bustling port metropolis off the Red Sea on the western edge of Saudi Arabia. My father looked around in awe. "*This*," he said, "is *ummah*."

Ummah? I didn't understand what this word meant. My father explained that it meant community. It's no wonder that I didn't know the meaning of this word. I had never felt a connection to the Muslim community in my hometown of Morgantown; they were so different from me in the way they lived their lives. With awe, I looked at the diversity in front of me. It was a window for me into the breadth of the Muslim ummah, and I was struck by its plurality. What I saw was people who were very different from each other coming together for a common purpose.

These particular pilgrims came from a place whose name comes from the Arabic *bilad al-sudan*, or "land of the blacks," where Arabic is the official language and Islam the official religion. In contrast to American Muslims, Muslims in Sudan live under political, religious, and social repression. Lieutenant General Umar Hasan Ahmad al-Bashir took power in 1989 in a military coup supported by the cleric-backed National Islamic Front. He oversaw the introduction of sharia law as well as the growing political and economic isolation of his country. But Sudan's large non-Arabic and non-Muslim population fought efforts to impose sharia nationwide. The country has been marked by civil war; tens of thousands have died, and Sudanese continue to die from war, famine, and genocide. Human rights activists allege one of the chief clients of a slave trade from Sudan is Saudi Arabia; Saudi Arabia and Sudan deny the charges.

My father, who knew all of this, was usually depressed by the abysmal state of Muslim leadership in the world. But he was hopeful for a respite from human troubles during our journey into a holy land. "To me, it's not just Saudi Arabia," my father said as we wove our way through terminal 4 of the Hajj Terminal. "It's the land where the prophet Muhammad walked."

To walk in his footsteps, about two million of the world's estimated 1.2 billion Muslims were expected to enter Saudi Arabia during this hajj. At one-fifth of the world population of six billion, Muslims are the second-largest faith community after Christians, who number about two billion. Because of high birthrates and conversions, Islam is widely considered to be the fastest-growing religion in the world. Although Islam is rooted in the Arab world, fewer than 15 percent of the world's Muslims are Arab, and half of the world's Muslims live in South Asia and Southeast Asia. The countries with the largest Muslim populations are Indonesia, Pakistan, Bangladesh, India, Egypt, Turkey, Iran, and Nigeria. The men and women in our tour group reflected this diversity. Although they now lived in America, they had come from Sudan, Palestine, Jordan, India, Pakistan, Afghanistan, Turkey, and other countries. The prophet said that the ummah has a serious responsibility to be nurturing and to represent Islam well in the world. I had long wondered if I could find a sense of community in my Muslim ummah.

On the hajj we were equalized by what we wore: the men were cloaked in the same seamless white fabrics, and the women in simple clothes. My father and Samir shuffled their feet beneath this uniform, called an *ihram*, or "restricted clothes." They changed into it when we stopped in Amman, Jordan, on the journey to Saudi Arabia. They wore one piece of seamless terrycloth wrapped around their midriffs to cover their bodies from the ankles to just above the belly button. They draped the other piece around their shoulders to cover the upper body. Straps can't be used to join the pieces together. Samir shuffled through the airport, looking like a little man, worried with each shuffle that his towel would slip from its precarious perch around his waist. He didn't want to go down in history as the first pilgrim to moon fellow hajjis. My mother and I had discarded our modern clothes and ornaments of our daily lives and now wore hajj uniforms that were like 1950s suburban Americana nightgowns with nun's habits. This modest uniformity marked our symbolic transition into our new roles as pilgrims and was also supposed to symbolize our entry into a state of purity and spirituality. We were wearing white, the color of the shrouds that are supposed to cover us when we are buried as Muslims. White is the color of death in Islam. A group of elderly Afghan women from our tour group looked like hard-bitten angels in their immaculate white gowns.

This uniform reflected a Buddhist concept I had come to understand during my travels: non-attachment. In the summer of 2000, during a weekend retreat at a forest monastery in West Virginia, the Bhavana Society, a Sri Lankan Buddhist monk named the Venerable Henepola Gunaratana had taught me the concept of mindfulness honed through the practice of a brand of meditation called Vipassana, meaning "insight" in the Pali language. "We learn to watch the arising of thought and perception with a feeling of serene detachment," he wrote in his book *Mindfulness in Plain English*. "We learn to view our own reactions to stimuli with calm and clarity." In a great hall at the monastery, with bullfrogs croaking outside in a lily pond and my eyes closed, I learned to view my thoughts as clouds passing overhead as I meditated. "This escape from the obsessive nature of thought produces a whole new view of reality. It is a complete paradigm shift, a total change in the perceptual mechanism. It brings with it the bliss of emancipation from obsessions." Islam calls this state of being *zuhd*, the Arabic word for an absence of attachment to matters of the world, or *dunya*. Zuhd is not an abrogation of responsibilities, but rather a deeper, more spiritual state of existence.

Around me, pilgrims were coming from the far corners of the world. The hajj is so important a journey that countries such as Indonesia and India, like Saudi Arabia, have a special area of their airports that is designated the "Hajj Terminal." For our hajj, a Muslim insurgent in Kashmir had personally lobbied India's prime minister, his enemy, to intervene to issue him a visa to perform the hajj. The king and queen of Malaysia had sent off their pilgrims in a formal ceremony.

But when I looked around I was stunned. Poor people. Rich people. Skinny people. Fat people. We were all there together. All the men, except the airport officials, were in the sacred togas of the hajj. All the women were in nightgownlike robes. But we were all unique. Just these few hours at the Jeddah airport had expanded my understanding of the Muslim identity. I had grown up seeing Muslims as mostly brown-skinned aunties and uncles and cousins from South Asia. In my teenage years, I started knowing them as fairer-skinned Arabs. Occasionally I had met African American Muslims and Muslims from Turkey and Malaysia. For the first time, I was now among them.

One of the group leaders told us that a lot would go wrong—and a lot would go right. In both circumstances, our leaders said, there was only one comment to make: *Al-hamdulillah*. "Praise be to God." Royal Jordanian lost our luggage. Al-hamdulillah. A woman tied Samir's shoelaces. Al-hamdulillah. Still hopeful, my father scoured the baggage carousel.

So much here was foreign. Between the baggage carousel and the passport counter, the sign marking the women's restroom depicted a woman with veiled gauze over her face. The man depicted in the sign for the men's restroom wore a beard. Here in Saudi Arabia, Wahhabism meant practicing the most puritanical interpretations of the Qur'an and the Sunnah ("path"), which includes the reported deeds, words, and silent approvals of the prophet Muhammad. Because the prophet wore a beard, Muslim men are supposed to wear beards, according to the logic of Wahhabism.

Terminal 4 is part of a massive complex with vast open spaces surrounded on all sides by terminals. I saw rows of women in *nikab*, the long black gowns and veils that women and girls are required to wear in public, under canopies that rose into the darkened sky. I felt strange about the fact that having had Shibli as an unmarried woman made me a criminal here, and that my son was evidence against me. We stepped up to the immigration counter, and an immigration officer who looked as if he was in his twenties made me pause. The man passed Safiyyah's passport to her with a twinkle in his eye and then, playing cat-and-mouse with her, pulled it away as she reached for it. He smiled. She didn't. I respected young Safiyyah. She didn't smile just to please this adult. I wished I had her sense of self-containment. I was leery of men like him in Saudi Arabia. Did they see Safiyyah, a girl on the cusp of womanhood, with less than innocent eyes? Or was he just innocent?

I thought I was going to be disgusted being in Saudi Arabia, but ironically I felt just fine. In fact, I kind of liked it. I walked freely through the airport. I certainly didn't feel like a pariah because I was a woman. Waiting for our bus to Mecca, we sat at a table at Al Baiader National Restaurant with Pepsi signs splashed by the restaurant's name.

My mother looked around frantically all of a sudden. "Where's Dada?" she yelled, worried. (My niece and nephew called their grandfather Dada, which means "paternal grandfather" in Urdu. They called my mother Dadi, which means "paternal grandmother.") My father turned up as quickly as he had disappeared. He was looking for our lost suitcase. He couldn't find it. Al-hamdulillah. Praise be to God.

Everywhere we turned there were packs of pilgrims from every country imaginable. A sign for pilgrims from the Philippines proclaimed, WELCOME

PILGRIMS. Another directed pilgrims from India in the direction we were walking. A dark-skinned man in bare feet stood near us. "Hajj Committee Bihar," my father read from his bag. Bihar is one of India's poorest states. The joke in India goes that Pakistan could have the disputed Indian-controlled state of Kashmir if it took the state of Bihar with it as well.

Everywhere, the Pepsi logo stared at us, from tablecloths to banners. We gathered with our tour group, piled against our suitcases between rows of shopping stalls. There were so many inside tricks to pulling off the hajj. A Saudi vendor pulled a belt from a bin of belts he was selling and tightened it around the lower half of Samir's ihram. Samir seemed visibly relieved.

The call for the predawn prayer rang through the air. In a restroom, I found myself around the largest sink I'd ever seen, with women from all over the world: Iran, Turkey, Indonesia. It was lower than normal, so that we could comfortably prop our feet under the faucet, and ringed with a low wall on which we could sit. We were all there to do precisely the same thing: wash ourselves for the ritual *wudu,* or "ablution." My mother taught me how to do wudu when she taught me to pray as a ten-year-old.

"Why do we have to do this before *every* single prayer?" I had whined, frustrated that I had to splash myself with water when I *seemed* clean.

My mother had looked at me, surprised that I would question a given that was tradition. But she humored my inquiry. "It's so that you will be clean before God," she said.

So I grew up privy to the universal code of ritualistic washing in Islam. Islam is so precise: the purity of having done wudu stays as long as you don't pass gas or go to the bathroom. I had to admit, however, that I retained my girlhood skepticism. Still, I washed myself before prayers, if not with conviction about the necessity or even the symbolism of the rite. Wash each hand three times, starting with the right hand. Wash the mouth three times with a gulp of water from my right hand to swish around in my mouth and spit out. Wash the nose three times with touches of water up the nose. Wash the face three times, from the top of my forehead to my chin. Wash up each arm to the elbow with a touch of water from the hand on the other arm, starting again, of course, with the right arm. Wash behind the ears three times with my fingers, both ears at the same time. Wipe the hair with my hands, three times. Run the backs of my fingers over both sides of my neck, intertwine my fingers behind my neck, and run them forward with the fronts of my fingers swiping my neck. This was my favorite move. Move right foot up to the faucet in an acrobatic move. Wash three times. It wasn't required, but my mother had taught me to run my fingers through my toes. I had always loved that part. Left foot and done.

It's said that the prophet Muhammad declared, "The key to heaven is prayer, and the key to prayer is being ritually pure."

I saw the woman beside me. She was shaking loose drops of water from her hand. She was from Iran. Even though we did not share the same language, we were privy to the same rites of ritualistic purity.

I stepped into the concourse to pray. The men were lining up in a massive space between shopping stalls crowded with glistening gold-colored water canteens, soft prayer rugs, and racks of *abaya*, the floor-length gowns that are the public uniform of Saudi women. I fell into prayer just behind them, a young woman directly beside me. For me, prayer was very much about going through the motions. I tried to get into it spiritually, but so many doubts interfered, not the least of which was why I had to pray *behind* the men and why my father *had* to be with me before I could enter this sacred space. I just hadn't fully surrendered to the faith. Still, there we all stood in the land of Islam, people of varying degrees of faith, all one in this concourse that was the Hajj Terminal, facing this effusive symbol of our faith, the Ka'bah.

I completed my prayer and lined up in a restaurant queue to get the family breakfast. The man behind the counter kept a lingering eye on me as I reeled off my order. "Half chicken with rice, chicken biryani, two apple juices, two mango juices, and two teas." I remembered the immigration officer playing cat-and-mouse with Safiyyah's passport. I had to admit that the idea of sexual tensions in this country gave me the creeps. As I walked back to our pile of suitcases, I caught sight of my mother. She was a walking sharia violation. As she scoured the stalls, her hair slipped out of her hijab. To my horror, she pulled the cover off in the middle of the crowd to adjust it.

"Mom!" I screamed. "We'll get arrested!"

I dragged her down into the suitcases with me to keep her from detection. Men and women chewed on wooden sticks, using them like brushes, as they shuffled by, their plastic sandals scuffing the tile with an irritating sound. "They're *miswak*," my father told me, explaining the sticks. In another hadith, or tradition of the prophet Muhammad, it was said that he brushed his teeth with the bark of a tree. It was another act called Sunnah, not obligatory but blessed. Hundreds of years later, strict devotees did the same to garner blessings, and the miswak was the toothbrush of these Muslims. It was another universal in the ummah, a tradition practiced by Muslims from villagers in India to men at my mosque in Morgantown. For my own dental care, I had packed an Oral B.

Jets rumbled overhead with more arriving pilgrims. A Bangladeshi man dragged a suitcase teetering on three wheels, the fourth wheel a casualty of the travel. A man walked by with a box on his head: HAJJ TOWELS. MADE IN CHINA. They were the terrycloth used in ihram like the kind Samir and my father wore. I snickered to my mother, "Made in China where no religion is allowed. Perfect."

As the sun started to rise, I caught a glimpse of the morning sky in an opening in the canopy. Soft, white clouds swept across a sea blue sky. The morning light cascaded into our holy space, warming the Muslims of the ummah who had assembled here, bathing everyone in a fresh beauty. *Maybe, just maybe,* I thought, *my family and I will be able do this journey without getting arrested.*

A young man approached Shibli. "As-salaam alaykum! How are you, youngest hajji?" He had delivered the Muslim greeting that means "Peace be upon you."

"Walaikum as salaam," I responded for Shibli. "And peace be upon you." Shibli looked at him with wide eyes and a curious smile. I appreciated this simple expression of kindness by my fellow Muslim brother. It reminded me of a moment while we were in transit in Jordan. A man in our group by the name of Hameed Omar pushed heavy airport chairs aside so I could squeeze Shibli's stroller close to a group leader giving a sermon. "You are not alone," he said to me quietly. "We will look out for you and help you." I looked at him and wondered if the Muslim ummah would indeed look out for me and help me. As a new mother raising her son alone, I found myself in the greatest position of need that I had ever known.

ON THE ROAD OF BIN LADEN

JEDDAH—I had to admit something: I was afraid for my safety. I was in a country that was totally defined by the repressive ideology that I was just learning about, Wahhabism.

When the Saudi royal family allied itself with Islamic evangelical Ibn 'Abd Al Wahhab in the eighteenth century, it was the beginning of the growth of a very puritanical branch of Islam. With the help of Lawrence of Arabia and the British, this alliance enabled the Saudis to remove Turkish rule from the Arabian Peninsula and the Middle East. Over the centuries the Saud family allowed the Wahhabi clerics to have control over the masses. It was their ideology that bred Osama bin Laden. He was a son of Saudi Arabia, and he used U.S. military presence in Saudi Arabia and Israel's control of most of Jerusalem to link his call for a holy war against the West to the rights of Muslims to maintain complete authority over the land where the three holiest mosques in Islam stand. Those three mosques were on our itinerary: the sacred mosque of Mecca; the sacred mosque in Medina, a city north of Mecca in Saudi Arabia; and a mosque called al-Aqsa in Jerusalem. Unfortunately, these sacred mosques are not just spiritual centers. They are also political symbols.

In an interview with CNN in 1997, bin Laden said the ongoing U.S. military presence in Saudi Arabia was an "occupation of the land of the holy places." In February 1998, bin Laden issued a fatwa, a religious ruling, calling for Muslims to kill Americans and their allies. Three other groups, including the Islamic Jihad in Egypt, endorsed the ruling. "The ruling to kill the Americans and their allies—civilians and military—is an individual duty for every Muslim who can do it in any country in which it is possible to do it, in order to liberate the al-Aqsa Mosque and the holy mosque [of Mecca] from their grip, and in order for their armies to move out of all the lands of Islam, defeated and unable to threaten any Muslim," read the statement, which was issued under the name of the "World Islamic Front." It was published three months later in the London newspaper *Al-Quds al-'Arabi*.

On the hajj, I stood in bin Laden's "lands of Islam."

He and the events of September 11 had made our religion a lightning rod. Saudi Arabia was the birthplace and breeding ground for most of the hijackers who flew planes that day into the World Trade Center and the Pentagon and crashed another plane into a Pennsylvania field. It is a country that has been skewered by Amnesty International and Human Rights Watch for its abuse of the human rights of dissidents, reformers, non-Muslims, and women. I stood in Saudi Arabia sad that my religion was being misrepresented by Osama bin Laden and his brand of puritanical Islam. No longer perceived in all their complexity and humanity, Muslims had become a monolithic enemy.

What troubled me even more was that our broader Muslim community was being taken over by right-wing Muslims. I'd seen it happen everywhere from my hometown in West Virginia to Pakistan, where Wahhabi ideology had taken root. This is the dilemma of all societies, including the United States, where moderate voices have been challenged by the emergence and increasing power of the religious right in politics. The handicap is obvious: extremists are usually more fanatical than moderates. On the hajj, our responsibility to the world was already becoming obvious to me: as moderates, we must be as impassioned about transforming the world with love and tolerance as the extremists of all faiths are about conquering it with hatred and division.

In thinking about fanaticism, I couldn't help but recall the connection between the fanaticism of Osama bin Laden and the development of modern-day Mecca and Medina. His father, Muhammad bin Laden, founded the Bin Laden Construction Group in the 1950s in Jeddah and since that time has spread his resulting wealth among his clan of fifty-four sons and daughters from several marriages. With close ties to the Saudi royal family, Muhammad bin Laden's company won a multibillion-dollar contract to expand Mecca and Medina. The construction firm's other big projects included building several royal palaces in Riyadh and Jeddah. It is no wonder that the mosques look like palaces: there is even a massive parking garage below the mosque in Medina. Following the Wahhabi edicts against anything that resembles worshiping the prophet Muhammad, the bin Laden empire paved over and dismantled many of the historical remnants of the original hajj. Osama bin Laden was disowned in 1994 when the Saudi government stripped him of his citizenship for his links to terrorism and his criticism of the al-Saud ruling family. But the family has remained in good standing despite its black sheep. Bin Laden Construction won the contract in 1998 to build a $150 million facility in al-Kharj, south of Riyadh, for the 4,300 U.S. troops based in the kingdom.

Kicked out of Saudi Arabia, Osama bin Laden sought refuge for his militant network first in Sudan and then in Afghanistan. After 9/11, America responded by trying to destroy his empire. What troubled me as an American Muslim was that the leaders of both Islam and the American government had betrayed essential principles of human decency and Islamic religious teaching with their politics of power. In that way, the West and the United States were complicit in creating the extremists within Islam who had become their adversaries. In Afghanistan, the United States had helped to create the vacuum filled by bin Laden and the Taliban. With Pakistan's intelligence agencies, the United States trained Muslim mujahideen, or "freedom fighters," to fight the Soviet takeover of the country, but when the Soviets pulled out, the United States did little to rebuild the Afghan economy, repatriate the Afghan refugees who had fled Soviet occupation into Pakistan, or disarm militants. The result: a fresh breeding ground for the Taliban and bin Laden's militant brand of Islam.

In Iran, in a plan sanctioned by the Eisenhower administration, the United States joined British intelligence agencies in a complex covert

plan called Operation Ajax that led to the ouster of the country's popular
prime minister, Mohammad Mossadeq, in 1951 after he refused to yield
power over Iran's oil fields to the British. The operation installed the
Shah of Iran, who maintained cozy relations with the United States and
the British over his country's oil supplies until anti-monarchy forces over-
threw him in the Iranian Revolution of 1979, in part out of retaliation for
the West's heavy-handed role in Iranian domestic affairs. In March 2000,
Secretary of State Madeleine Albright expressed regret that Mossadeq
had been ousted a half-century earlier: "The Eisenhower administration
believed its actions were justified for strategic reasons. But the coup was
clearly a setback for Iran's political development and it is easy to see now
why many Iranians continue to resent this intervention by America." In
addition, America had been friends with Israel since its inception in
1948, without resolving the crisis created when the Palestinians lost their
homes. I had heard my father's frustrations with world politics since my
earliest days. "It is not a question of Islam or Christianity, East or West,
democracy, justice, or freedom," my father always told me. "It's a question
of power. It's a question of modern-day colonialism of countries for money
and natural resources and the corruption of Muslim governments betray-
ing their people."

Now, as I began my pilgrimage, America was on the brink of war with
yet another Muslim nation. As we had sat at JFK Airport before takeoff,
CNN reported that President Bush and British prime minister Tony Blair
had met that day over plans to launch a strike against Iraq's Saddam Hussein. The headline: "Showdown: Iraq."

I couldn't help but feel sad.

OPEN BORDERS, CLOSED DOORS

And proclaim the hajj among mankind.
They will come to thee on foot and [mounted] on every camel,
lean on account of journeys through deep and distant mountain highways.
                          "Al-Hajj" (The Pilgrimage),
                          Qur'an 22:22

ON THE ROAD FROM JEDDAH TO MECCA—Throughout the changing tides
of history, pilgrims had overcome their fears to venture onto this path on
which I found myself.

Until the nineteenth century, a pilgrim usually traveled the long distance to Mecca by joining a caravan. There were three main caravans: the Egyptian caravan, which formed in Cairo; the Iraqi one, which set out from Baghdad; and the Syrian caravan, which, after 1453, started at Istanbul, gathered pilgrims along the way, and proceeded to Mecca from Damascus. Because the hajj journey took months, pilgrims carried the provisions they needed to sustain them on their trip. The caravans were elaborately supplied with amenities and security for rich pilgrims, but the poor often ran out of provisions and had to interrupt their journey in order to work and save up their earnings before they could continue. As a result, the hajj was a long journey of ten years or more for some pilgrims.

Travel in earlier days was filled with adventure. The Begum of Bhopal, a woman ruler from India, risked death to become the first royal pilgrim from India centuries ago. The roads were often unsafe owing to bandit raids. The terrain the pilgrims passed through was also dangerous, and natural hazards and diseases often claimed many lives along the way. For this reason, the safe return of pilgrims to their families was the occasion of joyous celebration and thanksgiving, a tradition that continues to this day.

With the days of caravans over, our modern-day pilgrimage with the Islamic Society of North America promised us air-conditioned buses, the Mecca Sheraton, and buffets. When we piled into our air-conditioned diesel bus at the Jeddah airport, we could sit freely wherever we wanted. On the bus there was no segregation of men and women, just as in my earliest days when I rode the yellow school bus that picked me up at the corner of Headlee and Briarwood Streets, a block from my house in Morgantown. I still preferred to sit in the back of the bus, and my family and I nested in the last row of our tour bus. My father sat next to my mother, and in front of us husbands sat next to wives. There was no men's bus or women's bus. I was surprised. I knew it was illegal for men and women who weren't married to mix freely in this country. This arrangement most certainly broke the rules. We hadn't even split the bus into a men's half and a women's half. On public buses in Pakistan, women had a small section separated from the larger men's section by a floor-to-ceiling metal screen. It seemed to me that the segregation only created a hypersexual society. When I was eighteen and in Pakistan for the first time, a man poked his finger through the screen into my rib just to feel a woman.

"Sicko!" I yelled, scowling, while my older cousin berated him.

Sitting in the back of our tour bus, with Shibli nestled against me, I felt conflicted as we went deeper into the land of Saud. I faced a contradiction. So often when people are faced with contradictions, they don't resolve them. I was trying to resolve mine, and even though this journey was risky, I knew it was the right thing for me to be doing. I felt safe in the refuge of our air-conditioned bus as we went deeper into this holy land. But I was also afraid, because Saudi Arabia is so notoriously repressive. Despite the risks, I was happy that I was there. One way to resolve contradictions is to create a delicate balance between safety and risk-taking.

To control the pilgrimage, the Saudi government required that each pilgrim be part of a group to which it assigned a Saudi monitor. In "Group A" of our tour group, we were given bright yellow wristbands bearing Arabic script we couldn't even read. They supposedly detailed the Saudi tour operator who had responsibility for our tour. Indeed, all sorts of politics expressed themselves during the hajj. Russia said police in Mecca arrested fifty-nine pilgrims from the Russian republic of Dagestan during the hajj in 2002 for trying to sell weapons, including rifles, swords, and night-vision goggles, to fellow Muslims.

On the road, lonely trees greeted us on a parched land spotted with trash. We were traveling forty-five miles east, inland to Mecca. For us, the road to Mecca meant traveling Spine Road No. 6. Gazing out the window, I thought about how different America and this land felt to me. I'd always loved the vast green spaces of America's East Coast. The land here was desolate. A sign told us we were on the road to "Makkah." We didn't even spell the names of cities the same way. Batik fabrics lay strewn out in the sun to dry. A factory filled the vista with concrete pylons reaching into the sky. Squat apartment buildings under construction dotted the landscape. We passed a billboard sign advertising Dunlop tires. Despite my travels, I was always surprised to find touchstones from our global economy in foreign lands. Two years before, I had stood in line at a McDonald's in New Delhi just to see what the Big Mac tasted like in India, since the fast-food chain respected the Hindu ban on eating beef and made its hamburgers with lamb. When I sank my mouth into the lamb Big Mac, I retched. It wasn't the same Big Mac I'd grown up eating in America.

Before we arrived in Mecca, we were supposed to declare our *niyyah*, or "intention," to perform hajj. I quietly declared my intention somewhere on the road between a billboard for DANON CREME CARAMEL and a sign for THE BEAUTIFUL CREATURES ZOO.

Lumbering down the road, we passed a sign for "Palestine St." and then a sign for "Falasteen St.," as "Palestine" is pronounced in Arabic. There was a sign for Ikea. Could it be the same as the one where I shopped for tea light candles and inexpensive furniture after graduate school? It was. Off the road from the airport, the Swedish home furnishings outlet had opened Ikea Jeddah in 1983 with its trademark Swedish meals in a restaurant inside; its hours of operation respected the Muslim custom of observing Friday as the Sabbath; the store opened after 5:00 P.M. that day. We passed a turnoff for King Abdul Aziz University, a sprawling university named after the country's founder. A chipped CD dangled from the rearview mirror of a truck passing by us. It was the same kind of CD with a Qur'anic verse printed on it that taxi drivers hang from their rearview mirrors in New York City for protection.

An eerie feeling consumed me. I felt as if the land had been raped and trashed. "This country haunts me," I wrote in my notebook.

We were headed into the Arabian Hijaz, a region of western Saudi Arabia that includes Mecca and Medina. I gazed at the tar road stretching in front of me. Using the back row of our bus as a refuge from the outside, I unbuttoned my shirt. After nursing, Shibli splayed out on my arms, milk-drunk. Shrubs dotted the landscape. I stared at the white pearl on the end of the stickpin tucked into the black head scarf of the young woman two rows in front of me. In the row in front of me, my mother stared out the window. She wouldn't have been there if it hadn't been for me.

"Look at this," my mother said, gazing out the window. "How many rocks there are."

"The better to stone people with, I guess," I said, under my breath.

Sometimes it was just easier to joke about the most oppressive interpretations of sharia. I tried to imagine the prophet Muhammad here on camelback. Someone propped a tire up with stones on both sides, like a modern sculpture. A man stood with a herd of sheep. I loved such scenes outside America. The United States was so sterile in comparison to most of the world. I had studied Arabic for two years at West Virginia University, but I could hardly read more than the "Allah" on the highway signs that we passed.

Allah was mentioned everywhere here in the birthplace of Islam. I remembered the Hindu pilgrims who had passed by me in buses marked "ShivShakti" when I visited goddess temples in India's Himalayan foothills. To them, the divine was expressed in Hindu deities such as the goddess Shakti and the god Shiva. I remembered the Buddhist pilgrims in the Tibetan pilgrimage in India. They meditated on the image of Tara, a goddess of compassion. What separated those faithful from the ones who filled my bus as it barreled toward Mecca? Our rituals, I believed, but not our core principles. We all believed in a higher being. We had all learned the golden rule: be kind, honest, and virtuous.

I thought about my mother's skepticism about the sacred worth of this land into which we were going deeper and deeper. "She doesn't believe," I wrote in my notebook.

Some hours into our journey I saw a tollbooth that led into Mecca and knew then why my mother didn't believe in the values perpetuated in this country in the name of Islam. In English, the sign read bold and clear: NO ENTRY FOR NON-MOSLEMS , using one of the transliterations of the word *Muslim*. The government knew how to be pluralistic about one thing: it translated the message into Japanese, French, and three other languages too blurry to recognize as the bus passed by. A sign over one of the lanes to the tollbooth spelled out the exclusive path on which we found ourselves:

MUSLIMS ONLY

↓

The other lane led to an exit ramp for non-Muslims. This made me sick to my stomach. For me, gaining entry into the Buddhist and Hindu pilgrimages had allowed me to understand—and appreciate—these two religions that otherwise would have remained mysterious to me. On those pilgrimages, I watched faith unfold before me in rituals, prostrations, and prayers that were strange to me. I heard the call to prayer in Sanskrit, Tibetan, and Hindi. I saw the devoted surrender themselves to a higher being with the hope of somehow improving their station on earth. I wouldn't accept having Rome, Jerusalem, or Allahabad, India, closed to me. I didn't believe in closing the doors of any community to others.

Mecca hasn't always been closed. For hundreds of years people of all tribes wove their caravans into the desert city of Mecca, sitting at a crossroads between the Western empire and the riches of the East. Mecca was a religious center at the time, and thousands of pilgrims each year would pay homage to hundreds of gods and goddesses enshrined in the giant black structure called the Ka'bah. To me, it wasn't wise to close Mecca to non-Muslims. We could only benefit by opening the doors of our Muslim communities to others, in the spirit of tolerance that we wanted others to show our community.

The Saudi economy had relied on foreign workers since the earliest days of the country's existence. U.S. oil companies led economic development in the country. And especially during the oil boom years of the 1970s and early 1980s, the state had used oil revenues to fund big development projects designed by foreign contractors, employing huge foreign workforces. There are fewer foreign workers today than during that period, but according to estimates, there are still more than 5 million expatriate workers in a country with a population of 24 million. Saudi Arabia's population is so young—45 percent are under the age of fifteen—that foreign workers are estimated to account for more than half of the total workforce. The economy relies on these workers, but the non-Muslim among them aren't allowed to step into the sacred space. Clearly, you can't be a non-Muslim Ikea delivery man entering Mecca.

It is the Wahhabi extremists like Osama bin Laden who want the doors to Islam's holiest places shut; one of his protests, after all, is that U.S. troops are stationed in the holy land of Islam. It seems to me the House of Saud would do Muslims a favor by opening the sacred city of Mecca to those who don't practice Islam.

I stared ahead at the exit ramp. Part of me wanted to escape. I felt a conflict in the fact that Saudi Arabia opened its borders to Ikea, foreign workers, and Western products but closed the doors of its holiest city. The Saudis had absorbed so many aspects of modernity, but their prejudices remained unchanged. I continued, however, on the road into Mecca. I felt there was no exit open to me.

HOUSE OF SAUD, HOUSE OF DONUTS

MECCA—We crossed into that place where most of the world's population cannot enter. We passed the sign that marked our entry into this sacred zone. HARAM BOUNDARY, it read.

For me, *haram*, which means "forbidden," has a negative connotation. It's used to characterize all actions that aren't considered Islamic. I found out later that a linguistically related word, pronounced *harram*, means "sacred" or "noble" in Arabic. I found that curious. But to me this place might as well have been called "forbidden." Most of my friends could never have come there. With a slightly different pronunciation, "sacred" becomes "sanctuary." We were headed into a space of the world called Haram Sharif, a "noble sanctuary."

Mecca sits in a narrow, sandy valley called the Valley of Abraham for the prophet of Judaism, Christianity, and Islam. The land is mostly barren with a rugged, rocky terrain and mountain ranges on the west, south, and east. It sits just 909 feet above sea level. The first thing I spotted as we crossed into this sacred space was a billboard for the exclusive French hotel chain the Sofitel. "*Mubarak*," I said to my mother, using the Arabic phrase—"May Allah bless you"—meant to communicate congratulations in my native India. "You made it to Mecca!" After *salaam*, the Muslim greeting of peace, *mubarak* was the only other insider Muslim language that I used, and slightly tongue-in-cheek at that. Arabic isn't my language, and I don't subscribe to the logic of those who want to declare Arabic the language of all Muslims. Many Muslims think Arabic is the language of God. We memorize the Qur'an in Arabic, as I did as a child, to internalize the word of God. I have often seen proficiency in Arabic used as a litmus test for how Muslim a person can claim to be. I inherited my own linguistic sensibility from my mother, who came from a literary family of poets and writers of Arabic, Persian, and Urdu. She didn't accept the inferiority complex that some people attach to a lack of fluency in Arabic.

On the road, an exit for the Inter-Continental Hotel veered off the highway. Acres of buildings loomed ahead of us, set on rocky hills, and the early afternoon sun was high in the sky. Sitting in the row in front of me, my father sighed with joy. "It's Mecca!"

We passed men scurrying along on the sidewalks. I didn't see one woman among them. We passed a billboard for Rado watches and an advertisement for Toshiba computers. Sprinklers shot water into the air on the side of the road. We hit a business district with storefronts advertising Sealy mattresses, Pepsi, and Farnas Rent-A-Car. I had so wondered what it would feel like to enter Mecca. It felt quite familiar. I could have been entering any other urban capital of the world. It was a mix of traditional ways and Western trappings. The manifestations of the modern economy were new creations. The shade of the trees and the stir of their leaves were the constants in life.

I spotted the first woman I'd seen on the streets. She was cloaked in black, her face crinkled. A store called Neha Optical advertised tinted contact lenses; behind the scenes, it seemed, vanity was universal. We passed a band of African pilgrims. They looked so regal and beautiful. One woman balanced a folded prayer rug on her head. We entered a tunnel that reminded me of the Holland Tunnel, which I'd taken as a child from New Jersey into Manhattan, only this entryway into Mecca had sidewalks for foot traffic.

Mecca was a plethora of sights as we emerged from the tunnel to busy streets with endless strips of stores and streams of pilgrims. A lone woman stepped down the steps leaving Sons of Saleh Musa Money Exchangers, which ran the lucrative business of exchanging foreign currencies for the local currency, riyal. She was a curiosity to me, someone to wonder about as I tried to understand how women lived their lives in this country.

The back of a bus carried a sign urging pilgrims to avoid the type of tragedy that ripped through tent colonies when lit cigarettes caused fires: TOGETHER. HAJJ WITHOUT TOBACCO. Vendors spilled onto the sidewalks, taking most of the space away from pedestrians. Gender dynamics were fascinating to watch here. A woman pilgrim held the strap of a man's shoulder bag, not daring to hold his hand. Couples rarely hold hands in Saudi Arabia and other traditional Muslim cultures. Even in America, I never saw my parents hold hands. They lived by the rules their parents had taught them for appropriate behavior between a man and a woman: no public displays of affection. We turned down a crowded, narrow street when I spotted an unexpected sign.

"House of Donuts!" I shouted. Next to the donut sign looming in front of us I saw a touchstone of my life in America, a familiar red-and-white-striped sign. "Kentucky Fried Chicken!" I exclaimed as my niece and nephew turned to look. "But where is Colonel Sanders?" Colonel Sanders is nowhere to be seen in Mecca. His goatee makes him acceptable in this land where wearing a beard is considered a signal of piety because the prophet Muhammad had a beard. But, strictly speaking, photography is illegal in Saudi Arabia. Creation is an act accorded only to God.

Just then we pulled in front of glimmering glass doors that led into the fourteen-story Sheraton Makkah Hotel and Towers, towering above the street. As we stepped into the gleaming marble lobby of the Sheraton and ascended from the elevator into the reception area, I saw that in the land of Saud there seemed to be an exception to every rule for those in positions of privilege. I looked up to see three larger-than-life images of the king and two princes of the House of Saud. To give a nod to the rule, the royalty didn't stare directly into the camera. It's a fundamentalist Muslim Kodak thing. The Taliban banned photography in Afghanistan, but even there leaders would get photographed, avoiding the camera's eye. I'd noticed that al-Qaeda upholds this practice, though Osama bin Laden often gives the camera the straight eye.

We went upstairs to our home base for this leg of our pilgrimage, room 708. It was an ordinary hotel room with two double beds and a TV, just like one of the many hotel rooms I'd stayed in on reporting trips from Long Island, New York, to Los Angeles, California. There was one notable difference, however: an arrow on the ceiling marked *qiblah* for the one direction in which we were supposed to keep ourselves focused, the Ka'bah.

CLIMAX AND ANTICLIMAX

The Kingdom of God is within you.
                                    Jesus (Luke 17:21)

MECCA—Past midnight, my father, my nephew, Shibli, and I slipped out of room 708 of the Mecca Sheraton with great anticipation. My niece wasn't feeling well, and my mother stayed behind with her, caretaking again. It was as my mother predicted. We were in Mecca, and she was in a hotel.

We were about to go into the most sacred of places—the Masjid al Haram, or "the Sacred Mosque," in which sat the Holy Grail of Islam—the Ka'bah. The Qur'an established it as a house of God. "Remember We made the House a place of assembly for men and a place of safety." Just to be particularly safe, we were going in the middle of the night to avoid the crowds that swarm the Ka'bah by day.

I was wearing layers of loose clothing for the required modesty. Underneath, I wore a long white polyester chemise my mother had bought at the Jeddah airport in one of the many stalls set up by entrepreneurs cashing in on pilgrims who didn't pack quite enough slippers, head scarves, or prayer rugs. Over that I wore a long white skirt my mother had bought at Kassar's, a grocery store and restaurant run by a kind Syrian family in Morgantown. On top I wore a kurta, a pink shirt from my mother's boutique on High Street in Morgantown; popular in the 1970s among hippies, this one was a new-millennium reintroduction at the Gap, with short slits on the seams that made it fit generously over my butt and hips. As a first layer on my head, I wore a tight scarf like the tube tops in fashion in the 1970s in America—only I used this one to keep strands of hair, not breasts, in place. Over that, and draping over me from my shoulders to my bottom, I wore a flowing white head scarf; it was the most convenient hijab I'd ever worn. In the middle it had a hole that I slipped over my face. It was just small enough to catch my chin and forehead in a very snug fit around my face. Atop my heart, my baby Shibli rested in his Baby Bjorn.

I had only seen images of the Ka'bah, embroidered, painted, stenciled, and replicated in every imaginable form. Physically the Ka'bah is an ordinary cube-shaped building, the size of a house, but it's a powerful symbol in Islam. It is the figurative house of God. We aren't supposed to worship the Ka'bah, but to focus on it.

On our way to stand before it, we stopped in the lobby for our group to assemble. Our guide was a young, unassuming man by the name of Sheikh Muhammad Alshareef. Born in 1975, he grew up in Canada and connected easily with all of us, whatever our generation. He reached my father with quotations from the Qur'an, chatted with Samir about Play Station 2, and won me over by leading us through stretching exercises like the kind I'd learned from 1970s marathon champion Bill Rodgers in old issues of *Runner's World*. Our sheikh had a serious air that offset his youth. He had graduated with a degree in Islamic law from the Islamic University of Madinah (as it's spelled there) in the class of 1999. He was the image of a pious Muslim with a full beard and gentle voice.

I pronounced the Arabic *sheikh* like the English word "shake," and I couldn't resist a pun. "It's Shake-and-Bake," I murmured to my nephew, not to be disrespectful but because the concept of a sheikh was so inaccessible to my irreverent American mind. A sheikh in Islam is sort of like a CEO in corporate America: instead of an MBA, a sheikh has a degree from an Islamic university or an Islamic scholar. It's an artful term for the leader of a group. Sheikh Alshareef passed out snappy business cards that identified him as executive director of AlMaghrib Institute, which ran classes on Islam accredited by places like al-Azhar University, a preeminent university in the Islamic world.

We eased outside. Even though it was late, the street was lit up with open storefronts. The House of Donuts was open. We slipped into a gentle wave of pilgrims streaming in one direction—to the Ka'bah. There was a calm buzz in the air as hundreds of other pilgrims walked with us to the Ka'bah. We passed the Kentucky Fried Chicken. As we walked we were supposed to chant a prayer called the *talbiya* to respond to God's call to us.

Labayk! (Here I am at your service!)
Allahumma labayk. (At Your service, oh Lord.)
Labayk. (Here I am at your service.)
La shareeka laka. (No partner do you have.)
Labayk. (Here I come.)
Innal hamda wan ni'mata. (Praise indeed and blessings are yours.)
Laka wal mulk. (And the dominion.)
La shareeka laka. (No partner do you have.)

It reminded me of the chant of a Hindu man whom I followed to a temple dedicated to a goddess in a corner of India. "Shakti Ma," he kept calling, beckoning the feminine goddess energy known as Shakti and associating it with *ma*, or "mother." It reminded me of the mantra, or the chant, of Buddhist monks as they circumambulated a special shrine in the city of Sarnath, outside Benares, India.

Like those faithful, we were supposed to utter the prayer with a sincere heart. All around me, my ear snatched utterances of the phrase, some loud, some quiet. I said the words and kept stumbling at "La shareeka laka. Labayk." (No partner do you have. Here I come.) It was like a tongue twister, but I kept repeating the spirit of the prayer to myself in my heart: "Here I come. At Your service, oh Lord." I wondered what my service would be. I didn't consider myself particularly extraordinary, but I felt as if I was destined to make a difference in this world. I just didn't know how.

As we pressed through the crowds, we repeated the call of pilgrims from time immemorial. Venturing to the Ka'bah for the first time, as we walked in the footsteps of those pilgrims who'd come before, was like a mysterious adventure. The desert air was cool, and the crowds thin but present. To make sure I didn't lose my nephew Samir, I tied a white cotton cummerbund between my wrist and his. As the road came to an end, the Sacred Mosque loomed in front of us like a religious albatross. I stared at it glittering against the night sky and couldn't help but feel like I was entering Disney World. We entered through a gate that I tried to squeeze through unscathed in the crowd.

When my father did the umrah in 1967, he entered through a humble main gate over a sand floor. The present haram dated back to 1570. It formed a central quadrangle surrounded by stone walls. But the mosque was nothing like it used to be. The hand of the wealthy Saud family had transformed even the physical experience of Hajj. In 1988 King Fahad Bin Abdulaziz started "the project of the second Saudi expansion" of the mosque in Mecca; this was the multibillion-dollar project awarded to the bin Laden family. Sure enough, the gate my father had once known had been replaced with a regal entryway called King Fahad, with minarets that reached almost 300 feet into the sky. The sand had been covered by marble tile. (My father would have preferred the sand.) The Saudis opened a three-story building to make up a side of the new mosque. The area of the mosque had been increased to 88 acres, including a rooftop prayer area—almost five times the area of the White House grounds and almost as large as the Vatican. Its capacity was increased from 410,000 pilgrims to 733,000, about ten times the number of football fans who squeeze into the Louisiana Superdome for the Super Bowl. The expansion added 56 escalators and 13 stairwells to the mosque. There were 1,091 places to do wudu, the ritual washing that precedes prayer.

A series of gates, shorter but just as regal as the big gate, encircled the mosque. We entered through one of those gates, stepping into the wide expanse of a marble plaza that surrounded the mosque like a perimeter of wealth. It was crowded but not suffocating on this marble pavement, called El Mataf. My father was beside Samir, Shibli, and me. Just as in a documentary I once watched, tracking the dance of people, men and women walked freely here, politely keeping some distance from each other and avoiding collisions. I was surprised at the freedom here. I could have been approaching the steps of the New York Public Library. I felt no inhibitions or restrictions as a woman approaching this daunting creation.

We slipped our sandals off and put them into plastic bags that Sheikh Alshareef had told us to bring to keep from misplacing them. We proceeded freely up a dozen steps to the doors that led into the actual mosque. As our sheikh led the way, I followed him. There was no women's entrance, as in the King Faisal mosque I'd entered in Islamabad, Pakistan, or my mosque back in Morgantown. There was no distinction of space separating men and women. We were one and the same here. It felt so liberating.

The Ka'bah sat inside a courtyard beyond the prayer halls that lined the inside of the mosque. A friend had told me that gliding through the passageways of the Sacred Mosque and emerging into the courtyard where the Ka'bah sits would be like going through the birth canal into the world. Around us in the massive inner halls, men and women moved in the different postures unique to prayer in Islam. Some were standing. Some were prostrating. Some were sitting. They were images of faith. As we walked my father recalled a story of devotion. There was a companion of the prophet, he told us, who had an arrow shot into his back during battle. As a reflection of the intense focus with which he prayed, he is said to have told his friends: "Take the arrow out while I am praying."

We continued through the massive halls. I sensed the light from the courtyard and started looking down at my feet. A physician in Islamabad told me that God will realize any prayer said at the precise moment that a pilgrim sees the Ka'bah for the first time. This physician was explaining her own spiritual path to me. I thought about her and the path of faith, surrender, and hope that brought devotees to Mecca. I had met her trying to get into Afghanistan. "When I looked at the Ka'bah," she told me as we zipped around Islamabad, "I asked Allah to make me a good Muslim." "You are either a Muslim or you are a *mu'min*," the physician said as she nosed her four-door car through the streets of Islamabad's upper-class neighborhoods. A mu'min is someone who is faithful. "My prayer came true. I became a mu'min." I didn't believe, however, in such hierarchical distinctions in people's faith.

In Mecca the moment arrived for me to see the Ka'bah for the first time. I was nervous. I was scared. I was excited. I was also cynical. Slowly, I opened my eyes and lifted them. I stepped into the light of the courtyard a little hopeful but so skeptical I couldn't take the pressure of a wish upon first sight. "The Ka'bah," I whispered to Shibli, whose young eyes focused on this place where history and faith intersected. Having safely brought my son here, I felt triumphant. I had heard stories of people weeping when they saw the Ka'bah, overwhelmed by the emotion of standing before this sacred image in Islam. I didn't see anyone outright weeping, but I saw the crowd getting worked up into a frenzy circling the square black box. For many people, it's a symbolic climax to their religious practice.

This building represented the first prophet Adam's connection to God, and thus our connection to God. It's said that after exiling Adam and Eve from paradise, God told Adam to build a shrine similar to one in the heavens known as Bait-ul-Ma'mur, the house in the seven heavens, where Islamic tradition says that seventy thousand angels circumambulate and worship Allah 24/7. The site chosen for the shrine on earth was Mecca. Adam built the building we now know as the Ka'bah. The story called it the first building on earth. The Qur'an says that Adam built the Ka'bah with the help of an angel, Jibril, who appears in the Old Testament and New Testament of Judaism and Christianity as Gabriel, a heavenly messenger of God's will. In Islam, Jibril brought a stone from paradise and embedded it in the eastern corner of the shrine. Other history says this stone was a meteorite. The stone was bright in the beginning, but with time it lost its luster and is now called Al-Hajar al-Aswad, "the Black Stone." When Adam finished building the house, the angel Jibril taught him the ceremonies of circumambulating the Ka'bah. It's said that the floods at the time of Noah destroyed the Ka'bah and only the black rock from heaven survived. Later, the prophet Abraham was dispatched to rebuild the Ka'bah.

Centuries later the job of rebuilding the Ka'bah was, ironically, passed on to the bin Laden family construction enterprise, and that political reality interfered with my surrender to the moment. Instead, what impressed me was the diversity I saw. There were people from all over the world in front of us. I heard the hum of different languages and absorbed the shades of different skin colors.

Most Muslims, like myself, spend their lifetime gazing at the image of the Ka'bah until they finally stand before it in reality. There is a funny tension between accepting the familiar in our lives and challenging ourselves with the unfamiliar. I tried to stretch my mind around the idea of this structure as a manifestation of the divine. For me, however, it was an anticlimactic moment. The Ka'bah itself wasn't attractive. Most of us are looking for symbols of God that bring us a sense of spiritual completion. Some of us in Islam find it at places like the Ka'bah. I didn't. What the Ka'bah said to me was that I was still searching. I even forgot to make a wish.

We dove into the sea of humanity around the Ka'bah and started our first circumambulation facing the black stone. We hadn't even started the actual hajj. Sort of like a practice round, we were doing umrah. It replicated the rituals in Mecca that we did on the hajj. We had to walk seven times around the Ka'bah—a ritual called *tawaf*, or "encircling"—reciting the call of pilgrims. Our circumambulations are supposed to reflect how our lives revolved around God.

As he circled the Ka'bah, the prophet Muhammad kept saying, "Our Lord, give us goodness in this world and goodness in the Hereafter, and keep us safe from the fire of Hell." I could only repeat the first half of the prayer. I just couldn't get motivated by fear of this concept called hell.

Men jogged around the Ka'bah in the first three rounds in a practice called *ramal.* Women weren't allowed to run, an edict I didn't like but wasn't about to defy. The runner in me wanted to break into full stride. The courtyard was packed with women in black nikab and batons. The nikab is the most hard-core covering that Muslim women use; here, it was black and shrouded women from head to toe in fabric and full-face veils with netting over their eyes. They looked like ninjas. In Afghanistan, under the Taliban and after, the women were shrouded in similar coverings, made up in blue fabric with veils styled slightly differently. I wasn't about to mess with them. One of them, a Saudi policewoman, came to life, playing with Shibli.

I saw the *burka,* or "veil," that was the curtain that covered the door of the Ka'bah. As we rounded the corner I eyed a green light and, standing near it, turned and faced the Ka'bah. So did, it seemed, just about everybody else. Shibli dangled his feet in the crowd. This was the point that marked the beginning of each round of the tawaf, and this was where we were supposed to do a ritual called *istilam*—kissing the Ka'bah, touching it, or simply facing it to honor its divine history.

Around me, pilgrims pressed hard up against each other trying to get to the stone. It's said that the prophet kissed the stone during his last hajj, so among pilgrims it's considered blessed to kiss the stone if possible, or to just give it a flying kiss otherwise. The tribe of Muhammad, the Quraysh, rebuilt the Ka'bah in the seventh century, and the prophet put the black stone into the structure with his own hands.

I tried to eye the stone, but in the crush of the crowd I couldn't see it. Anyway, I wasn't about to even symbolically express affection to this rock. To me, it represented mythology used to secure our physical connection to the divine. I only had to feel the softness of Shibli's fingers wrapped around my index finger to know that the chord was not cut. With this revelation, I was coming to terms with what I believed rather than simply embracing what I was *supposed* to believe. It seemed to me that all of us, Christians, Jews, Muslims, Hindus, and Buddhists, have to challenge our faiths. As I traveled in India I was told, often in adulating tones, about a Sanskrit concept of blind faith. I felt that our world and our religions would be better served by conscious, mindful faith.

It was madness near the Ka'bah as pilgrims threw themselves against its walls to try to kiss the stone. It took such faith and devotion to fling forward. The situation at the Ka'bah reminded me of the time I wiggled my way into the mosh pit at a No Doubt concert to sing along with Gwen Stefani, "I'm just a girl." I loved mosh pits, but I didn't get the scene in front of me. The journalist in me kicked into high gear.

Crazy, I whispered to myself, crushed by the press of pilgrims.

The frenzy was not very different from the rush that filled the air when I'd watched Buddhist pilgrims stampede the stairs of the Ki monastery in the Himalayan mountains of India just to set their eyes on a holy mandala, a circular creation of geometric designs that symbolize a blessed circle of protection. When I closed my eyes, I could see the dust storm kicked up by two hundred naked Hindu yogis, called *naga babas,* as they bolted for their holy ritual bathing in the Ganges River during the Maha Kumbha Mela. It was the same devotion that sent Jews and Christians to their pilgrimage sites. I had to admit that I didn't feel the surrender to my faith that I was told I should feel at a moment like that. I somehow wished that I could be like them. But I wasn't.

In this crowd, I felt as if I was going through the motions of something I couldn't fully understand. The real awe came in seeing the unity of a people in one act. The African men gliding through the inner circle held my attention. They were dark and muscular with glistening sweat trickling down their bare torsos. With four of them carrying one pilgrim, they held wooden stretchers overhead, and elderly pilgrims sat in a leisurely way against the stretchers' short walls, like pharaohs of yesterday. They bobbed so fiercely through the crowd that they virtually danced around the Ka'bah.

We wound around and around. At the end of the last round we found a corner in the courtyard, and I offered the requisite prayer at Maqam Ibrahim, or the "standing place of the prophet Abraham," marked by a small golden kiosk. This was the spot in Mecca toward which, in a world of about six billion people, about one in five people turned for their daily worship. Women from Indonesia walked briskly inside a protective phalanx of men. Men and women with the flag of Turkey sewn onto their jackets scampered by us, trying not to lose each other. Some of the women in black burkas seemed to wear visors beneath their veils. This convergence of humanity was amazing. A Turkish woman saw Shibli and smiled.

"Ma sha Allah!" she shouted. "This is according to Allah's will!"

Another woman, wearing a scarf printed with hearts, bounced in front of us and smiled at Shibli as I thought about this phrase. I had always thought that it was an expression of simple praise to God for anything that seemed attractive, desirable, or admirable. It was Muslim insider language to protect someone from the evil effects of envy. But its literal meaning had extra resonance for me. I believed Shibli's conception was God's will. I appreciated every utterance of this invocation because I was also very aware that I had to protect Shibli from the negativity of judgment in our community.

With the image from tapestries in front of me, I started rethinking the mythology that was passed down through the ages about the Ka'bah. I understood why outsiders looked at this rite and equated it with pagan worship. This devotion to the physical structure of the Ka'bah struck me as contradictory to Islam's teachings prohibiting idolatry.

It is true that before the prophet Muhammad started preaching Islam, pagans revered the black stone that was the Ka'bah. They circled the Ka'bah in white robes and called out the names of the pagan gods *and* goddesses. The chief pagan god was the god of Mecca and Ka'bah, Hubal, or "al-Lah," "the God." Our modern-day Muslim way of referring to God as Allah came from this ancient name, but the Qur'an doesn't even mention him. It talks about his three daughters: the goddesses al-Uzza, Manat, and, most significantly, al-Lat, the fertility goddess, or "the Goddess." Like most religions, Islam came from a pagan tradition that revered the power of a feminine divine. I had learned this while studying Tantra, a philosophy rooted in goddess worship. From Egypt to Babylonia, Greece, Rome, Asia, Africa, and the ancient cultures of the Americas, ancient people related to God in feminine as well as masculine terms. Some Jewish scholars and followers of the Kabbalah Jewish spiritual tradition even believe that Yahweh, the Hebrew God, can be traced to a goddess, Shekhina. Some historians say it is very likely that the Ka'bah was originally a source of astral worship, a common theme in goddess traditions. The symbol of Islam in the modern day—a crescent and star—captures the spirit of that early devotion to the heavens.

As we stood in front of the Ka'bah, Sheikh Alshareef read my mind. "It may seem weird to pray at the rock." It did. "The Ka'bah is the direction where we pray," he said. "That's all."

THE DIVINE IN THE DESERT

And God heard the voice of the lad;
And the angel of God called to Hagar out of heaven,
And said unto her, What aileth thee, Hagar?
Fear not; for God hath heard the voice of the lad where he is.
                                   Genesis 21:16–19

MECCA—There is a place in this sacred city that is even more important to me than the Ka'bah. It is a path between two hills where the most remarkable woman once ran in desperation, searching for water for her son. She is Hajar, or Hagar in the Bible and Jewish history. Her name, which means "to take flight" in Arabic, is the linguistic root for *hijrah*, but nowhere in Mecca can you see her name. I had never heard the story of Hajar until I started getting ready to go on the hajj. She is one of the forgotten heroines of Islam. Her life is overshadowed by the story of a man, Abraham.

The Old Testament story from Genesis says that Abraham could not father any children with his wife Sarah. On a trip to Egypt they bought a young slave woman and returned to their home in Palestine with her. She was Hajar. In Islamic history, Abraham married her as his second wife, a co-wife. Unable to have a child, Sarah told Abraham to have sex with Hajar so that he could have a child, making her an early surrogate mother. It was apparently a practice of that time. A son was born of this union between Abraham and Hajar, and Abraham named him Ishmael. Mothers in the Arab world got their identity from their children, and Hajar became Umm Ishmael, or "mother of Ishmael." (Abraham would have been "Abu Ishmael" or "father of Ishmael.") The tale continues like a script from a 2000 B.C. soap opera. Hajar's fertility tormented Sarah. The Qur'an doesn't speak about this rivalry, but maybe God could predict what would happen next in the story, having after all created human nature. Jealous of Hajar, Sarah ordered Abraham to banish the servant to the desert. Abraham complied.

According to the Qur'an, Allah ordered Abraham, in a test of faith, to take Hajar and Ishmael to the parched desert in the valley of Mecca, then called Bakkah. Hajar placed Ishmael on the same ground that now lies beneath the marble floors of the Sacred Mosque in Mecca. Abraham walked away from her after placing a bag of dates and a skin full of water beside Ishmael. Hajar turned to him and asked, "Abraham! Where are you going? Why have you left me in the wilderness where none is to take pity on us? Nothing is available here to eat and drink."

She kept repeating herself, but Abraham didn't listen. Then she asked him, "Has God commanded you to do so?"

He replied that God had indeed so commanded him. She protested no longer. "Then God will cause no harm to me." In Islamic history, Hajar made the choice to accept Abraham's decision. She could have clung to him. Instead, she chose to turn her back on Abraham and walk away from him. Clinging to faith in both God and herself, Hajar was the image of strength. Four thousand years ago, she was standing alone in Mecca.

According to the Qur'an, as Abraham left Hajar and his son, he said, "Oh Lord! I have made some of my offspring settle in this barren valley near the sacred house so that they may keep up prayer." When Abraham got out of Hajar's sight, he turned toward Ka'bah and prayed: "Oh Lord! Grant that the hearts of some men may be affected with kindness toward them and bestow upon them all sorts of fruit so that they may be thankful."

With those words, he left Hajar and Ishmael. He returned to the life he had built with Sarah. Hajar, meanwhile, struggled, like every mother, to give her child a good life. She was subjected to one of the most difficult trials God sent down to earth. In her place of isolation, Hajar began to suckle her child and drink water out of the skin Abraham had left them. Finally, the water ran dry and she ran dry. Ishmael started crying for milk. At the time of Hajar, there was no Ka'bah drawing millions of pilgrims to it every year. Desperate, Hajar ran seven times between this place called Safa and another hill called Marwah, searching for water.

As she ran she yelled, "Oh Lord, forgive, have mercy. Ignore our sins. Of course, You know what we know not—only You are the Holy, Merciful."

Hajar was about to start the eighth trip between Safa and Marwah when she collapsed next to Ishmael. Her eyes turned to her crying son, who was kicking the ground in agony from thirst. It's said that the angel Jibril, or Gabriel, caused water to spring forth from the earth where he kicked. Hajar saw the water oozing out of a hole in the ground near her child.

Seeing the precious water escaping into the surrounding sand, she cried, "Zumi, ya Mubaraka!" (Stop there, O Blessed water!)

A pool formed as she approached it, and this wellspring of holy water was from that moment called *zamzam*, meaning "to stop." Hajar drank the water, and Ishmael nursed from her, both their lives saved.

Through her strength of character, Hajar became mother to a new civilization. With Abraham building a life for himself with Sarah, Hajar raised Ishmael alone near the spring of water that had sprung up, an early single mother. One day some people from a tribe called Jurhum passed through. Seeing a bird that had the habit of staying near water, they sent a messenger, who discovered the source of the water in the zamzam spring. They became the first people to settle in the area after Hajar. Hajar arranged her son's marriage to a daughter of the tribe. According to Muslim lore, Abraham returned one day and interrogated the young woman chosen by Hajar. Disapproving of her, he sent a message to Ishmael that he should divorce her and marry the daughter of the tribe's chief. Ishmael complied. That union spawned the Quraysh, the Arab tribe into which the prophet Muhammad was born centuries later.

Meanwhile, in the city of Hebron outside Jerusalem, Sarah conceived a child with Abraham, a son named Isaac. Their son Isaac married Rebecca, and from them were spawned the Jewish tribes of the Middle East.

Hajar should have had a revered place in Islam. Instead, even her choice of a bride for the son she raised was rejected. She is not mentioned by name in the Qur'an. And the history books have always identified her as her son's mother. The prophet Muhammad said, "May Allah bestow mercy on Ishmael's mother!" He didn't mention her by name, but he did at least honor her. The credit for the hajj and the building of the Ka'bah goes to Abraham, the man who abandoned Hajar and Ishmael. Abraham returned to the desert, the Qur'an says, and with his son Ishmael rebuilt the Ka'bah, originally constructed by Adam but destroyed in the great floods that saw Noah's ark travel the world.

Remember when Abraham and Ishmael built the foundations of the House [and prayed], "Our Lord! Accept this from us because You are the Hearing and Knowing. Our Lord! Make us compliant people who bow to You. And of our descendants, make them compliant

Permanent Committee of Islamic Research and Fatwa" and Sheikh Muhammad Bin Saleh al-Uthaimin, and printed and distributed by "The Ministry of Islamic Affairs, Endowments, Dawah and Guidance, Kingdom of Saudi Arabia." *Dawah* means to invite others to Islam either literally or through preaching. It irritated me that books in Saudi Arabia communicated such a sense of moral authority, using words like *enlightenment* and *guidance* when they were nothing more than ideological polemics of a branch of Islam that was divisive, intolerant, and sexist. Without acknowledging any of the scholarly debate on interpretations of the Qur'an, the traditions of the prophet, and the sharia, the rulebook alleged that any who didn't believe in these precepts were nonbelievers destined for hell. The precepts were, conveniently, the tenets of Wahhabism, including face veils for women, the stoning sentence for adultery, the cutting off of hands for theft, and the classification of Muslims, not to mention non-Muslims, who didn't follow the rulebook as "unbelievers."

What especially offended me was that as the gatekeepers to this act of the pilgrimage—required of all Muslims—Saudi religious clerics had a captive audience and a specific window of opportunity in their mission to convert Muslims to their Wahhabi ideology. I had met so many Muslims who believed that if a point of view came out of Saudi Arabia it had to be theologically correct. After all, the logic went, the Saudis are direct descendants of the prophet and the closest in the Muslim world geographically to his teachings. So often in Pakistan after 9/11 I had heard educated Muslims proclaim that the Saudis and the Taliban were practicing the most authentic Islam in the world. That gullibility both saddened and horrified me. What I suspected was that in fact, as an ideology, the Wahhabi school had departed more than most from the original teachings of the prophet.

I closed the book, my finger marking the page, and set off alone with Shibli in the path of Hajar. At each hill I recited the prayer again. I walked briskly along these steps that Hajar had run in desperation. Green fluorescent lights rose from domes that marked each hill, reaching into the dark sky like strange beacons.

In another window into the hudud, or "boundaries," erected around women over the centuries, Saudi law prohibits women from running the distance of the walkway, even though Hajar herself ran in desperation to find water and men are allowed to sprint. This ban reflects the way in which men have redefined women's roles throughout history. The edict came from the days after the prophet Muhammad. Ibn Umar, one of the first leaders of Islam, said: "Women are not obligated to jog around the Ka'bah, nor while making sa'y between Safa and Marwah."

This admonition might have started as an allowance, not an order, to give women a physical break, but it had turned into a rigid rule. A little over a year before, in Pakistan, I had asked an aunt who had done the pilgrimage, "Why can't I run?"

"Just obey," she had answered. No questions allowed.

As I followed in Hajar's footsteps, my external reality was quite different from hers. I was traveling with a packet of Cottonelle Fresh Folded Wipes, premoistened and flushable. "Now with Vitamin E!" it said on the package. But my internal reality was very much in sync with what I knew about Hajar. I felt such compassionate empathy with her. I had made a choice, like her, to raise my son alone, contrary to the traditions of our cultures. Although millions of people swirled around me, I too felt as if I were standing alone in Mecca.

In today's society, Abraham's abandonment of mother and child in the desert would make him a deadbeat dad, sacrilegious as that sounds. He didn't wire Western Union child support payments to the desert. In my home state of West Virginia, social workers would have moved to dock his paycheck. I felt angry at Abraham and wondered if I could ever forgive him and feel compassion for him. Clearly, I was projecting my own emotional experience and cultural context onto history. I felt as if the biological father of my son was the Abraham in my life. Ironically, my house in Karachi, the place where I most likely conceived, was on a street named after the holy water that Hajar had found. I had lived on Zamzama Street. When I left that house, my unborn son within me, I was scared and confused, but I was determined to persevere, surrendering myself to divine will and my own resources.

In the path of Hajar, I repeated her words to myself. I prayed to God those words that had crossed Hajar's lips. "Have mercy," I prayed. "Lord, have mercy on my son and me."

It is our responsibility on this earth, the tradition goes, to respond to the call of God. Rumi, the mystical saint of centuries ago, once instructed his students to breathe in the divine of the universe with the utterance of Allah in *zikr*, or the remembrance of God. Breathe in the divine of the universe with the expression of *Allah*. Breathe out the inner divine into the universe, he said, with the utterance of *hu*, meaning "is." I sang to Shibli as I walked between Safa and Marwah, "Allah-hu! Allah-hu! Allah-hu! Allah-hu!" God is. The divine is.

The run, or sa'y, was meant to remind us of a mother's struggle in search of mercy. Broadly, it represents the struggle of each one of us for mercy. But what was Hajar's sin for which she needed mercy? It seems, if you believe Genesis, that she taunted Sarah, but maybe this young woman was merely seeking legitimacy from the father of her child. These days, her sa'y could have been a lifetime of the night shift behind the cash register at the local 7-Eleven, supporting herself and her baby. But in the ancient world it meant that she ran alone in the desert in a ritual I commemorated with Shibli.

I looked at my son. Shibli's birth had awakened in me a new mindfulness and clarity that I had never had before about my purpose in life. For most of my adult life, I was a woman trying to reconcile the dissonance between my birth, religion, and ancestry and my Western upbringing. I had wandered the world for two and a half years, from meditation caves off the coast of Thailand to grass huts on the banks of the Indus River. I had returned to Morgantown, betrayed, disillusioned, and yet somehow still hopeful. With his birth, Shibli had given me new life.

Islam encourages men and women to be courageous in their spiritual jihad. I tried to fulfill the expectations of my culture by marrying a man of my faith. When it became clear after I married him that we were incompatible, I had to overcome my cultural programming to leave him, believing that Allah did not want me to suffer in a suffocating, emotionally empty marriage. Ten years later I chose to leave the father of my baby after it became clear that the relationship would be unfulfilling and tumultuous. I realized that I had been trying to stay together with him for the wrong reasons. I felt incomplete, and I had wanted him to complete me. I felt illegitimate without him as my husband, and I wanted him to make me legitimate. With this understanding, I dedicated myself to centering myself on my own life. I wanted to offer Shibli the best I could provide. Throughout my life I had felt so much pressure from my culture and religion to get married that I often stayed in dead-end or unhealthy relationships longer than I should have. Breaking free of unhappy relationships was symbolic of my wider effort to question the rigid acceptance of rules set forth by others, be they religious, government, or community leaders or family members. Islam reinforced the idea that we must think for ourselves, and so did my common sense. I was trying very hard to do just that. Religion is supposed to inspire the best in ourselves. That was why, despite being abandoned by Shibli's father and despite the judgment of others, I could be in Mecca and stand with my child before God.

I had drawn on the best in myself to try to save my friend Danny. I drew on the best in myself to bring my son into the world. It was because I went through so much that I could hold my child without shame.

68STANDING ALONE

Men and women don't define what it means to be a good Muslim. It is defined by the core universal values of what it means be a good person. I went through that struggle. And that is how I became legitimate. The sincerity of Hajar's heart allowed her to find her zamzam. She had to tap the best of herself to find the water. She was alone. She was desperate. God knew her sincerity and answered her prayers. She didn't need any intermediaries.

We had slipped into the earliest hours of the morning, just before dawn. I was far from the earth where Hajar ran in desperation, looking for water, but I felt close to her spirit. With the heavens above me, it was as if I could feel the pulse of not only Hajar but every mother since the beginning of time.

I was praying without pause, making resolutions from my heart for a wise and good life with my son. Women continually kissed Shibli on his forehead. "Little hajji!" the women exclaimed. My heart smiled with each expression of love for Shibli. They didn't know that I had no wedding ring. A spiritual umbilical cord connects all women through the timeless universality of motherhood. I was one of those women.

A flurry of birds called ababeel dashed and darted above us in the night sky, as if to punctuate my celebration of motherhood. They squealed and chirped in a concert both eerie and lyrical. My father told me later that these birds hold a divine place in Islamic history. It's said that an Abyssinian leader named Abraha assembled an army of sixty thousand warriors and elephants to destroy the Ka'bah, coincidentally in the year the prophet was born. When the warrior reached a valley outside Mecca between two cities called Muzdalifah and Mina, the elephants knelt down and refused to proceed to Mecca. These birds pummeled the soldiers with rocks, forcing their defeat.

Have you not seen how your Lord dealt with the Companions of the
    Elephants?
Did He not make their treacherous plan go astray?
And He sent against them Flights of Birds,
Striking them with stones of baked clay.
And He made them like an empty field of stalks and straw of which the
    green crops had been eaten up.

"Al-Fil" (The Elephant),
Qur'an 14:105

Divine expressions are often used to glorify human destruction. In Iraq the Thou al-Faqar Factory in the city of al-Taji, north of Baghdad, produced missiles called Ababeel. In Pakistan the military used a low-speed target drone called Ababeel for antiaircraft gunnery training. And in Syria a journalist and human rights activist, Nizar Nayyouf, spoke about a system of torture, "the Stones of Ababeel," that he witnessed in 1993 while being held as a political prisoner in a penitentiary called Tadmor. From a height of about twenty feet, guards would drop concrete blocks on prisoners walking to the courtyard. The prisoners often died, their skulls crushed.

Ahead of me, I saw my father and Samir wrapped in their sacred togas. They were at polar opposites of manhood. Samir's towels flapped while he ran between the two green lights that marked Safa and Marwah. As they passed me on the return I saw tears in my father's eyes. He was thinking about Hajar and the struggle of this mother alone in the desert with her son. Samir looked at him, curiously. With the clarity of a wise child, he came up with a remedy for his grandfather's tears. "If you don't cry," he told my father, "you can pray louder."

My father smiled at him and dried his tears.

As we continued through the late hours of the night, Shibli stayed awake and alert. With both hands, I grasped his fingers around mine, gently extending his arms out like rudders steering us forward. Boys pushed wheelchair-bound pilgrims beside us, sprinting and then jumping onto the backs of the wheelchairs from time to time. I secured Shibli's body with my hands and tried to move more briskly without breaking into a full run, lest the religious police reprimand me. Back and forth we went. At each hill we said a prayer. I saw my father and Samir complete their rounds: my father pulled out a pair of scissors to snip a lock of Samir's hair, a symbolic practice to mark the end of this ritual.

After them, I completed my rounds just as Shibli started to stir restlessly. His cries started slowly but soon built into a plaintive wail. He was hungry. I looked around me. The walkway had become crowded. Crowds had arrived for the predawn prayer, called *fajr*; it was said that the closer you prayed to the Ka'bah, the more blessings you got. I didn't know how I could feed my hungry son in this crowd. I was a mother desperate to feed my son. His cries grew louder. I grew more desperate. I dropped to the floor facing a wall off the path of Hajar, sitting in the direction of the Ka'bah below me, straddling Shibli on my lap. I reached under the bottom of my hijab, which fell by my waist, and frantically began undoing the buttons on my long shirt, but my chemise was tight around my breasts. I turned to my father beside me and in desperation asked for his scissors. I cut the top of my chemise and ripped it open with my bare hands. I drew Shibli to me. His desperate lips found the milk within me, flowing to him like holy water onto his parched lips. I felt as if I was connected to Shibli with the eternal bond that linked Hajar to Ishmael. It was the life force of creation that touched everyone and everything around us, before us, and after us. To me, it was what we call God. It was what we call Allah.

As Shibli suckled, the sounds of the morning call to prayer erupted into the cool air and reverberated in his ears. It was the same azan, the same call to prayer, that my father had whispered into his ear at his birth. In the holiest place for Muslims, Shibli was hearing the call from the muezzin, the person who makes the call to prayer, of Mecca. There is only one line added to the morning azan: "Prayer is better than sleep."

That moment meant so much to me. I was in Mecca, a criminal in this land for having given birth without a wedding ring on my finger. And I was nursing my son at the holy mosque of Mecca, overlooking the sacred Ka'bah. This was nature's law expressing itself, more powerfully than man's law. I drank the sacred water called zamzam. From me, it flowed into Shibli. I recognized then the great lineage I had in Islam. I was a daughter of Hajar. I looked up to the sky with one thought: blessed are the daughters of Hajar.

A PRAYER SIDE BY SIDE

MECCA—As the call for prayer broke through the air, something drew me away. "Let's go," I told my father.

"Now?" he said.

"Yes, let's go," I urged him again.

He was an accommodating man, even though he wanted to pray there. He saw my brows furrowed. He saw I felt tense in the crowd. We tumbled through a rush of pilgrims jamming their way to enter into the Masjid al-Haram, the Sacred Mosque. We emerged onto the road on which the Sheraton sat, when I saw two familiar faces in the crowd: my mother and my niece, Safiyyah. We had prayed for Safiyyah's return to good health, and now she was bouncing again with youthful energy. "Your prayers worked," my mother exclaimed.

I felt complete, reunited with my family and standing together in Mecca. While we were standing there, more people spilled out onto the streets to line up for prayer. We decided to join the congregation near where we stood. We were not within the gates of the Sacred Mosque, but we were part of the congregation that spilled out of the Ka'bah. Prayer started, and we stood together, shoulder to shoulder as a family. We had prayed this way at home, but never before in congregation with other Muslims. One time when I visited our local mosque in Morgantown, I was separated from my father and found myself alone in the "women's section," isolated from a band of men giggling as they read the Qur'an aloud in Arabic, making fun of each other's mistakes in pronunciation. It was a different experience in Mecca. There were no formal boundaries between men and women, between boys and girls. Families prayed together. Men and women who happened to pray beside a stranger, as many of us did, tried to pray beside someone of the same gender, but it didn't always work out that way and nobody ruled mixed-gender prayer lines indecent. It was no more complicated than that. There were no curtains, walls, or partitions dividing men and women from each other, just common sense. In contrast to my mosque in Morgantown, this arrangement made me feel respected and valued. The idea that I needed a wall to protect me from the guiles of men seemed insulting. Here in Mecca the arrangement was more natural. This experience, in retrospect, planted the seed for actions I never thought I'd have the courage—or the will—to take toward reclaiming women's rights in Muslim communities.

We stood in line like this with at least another million pilgrims in front of us, behind us, and around us, all of us facing the Ka'bah. Near us, some men and women stood on cardboard boxes they had flattened to use as prayer mats. Somewhere in each of our personal histories, different as they were on so many accounts, we had all learned and memorized the movements that make up Muslim prayer. I was ten when my mother taught me how to pray at home. She guided me gently through the stages of prayer. When I did yoga as an adult, the yogic positions that were a part of Hinduism, a religion that predated Islam, would remind me of the postures I'd learned for Muslim prayer. The prostrations are meant to capture the surrender that defines Islam and the social justice and equality that are supposed to define Muslim communities. At the time of the prophet's revelation, the society around Mecca was stratified into the upper classes and lower classes that made up most societies. The prostrations he prescribed were meant to dismantle hardened egos and equalize everyone before God. Growing up with the experience of putting my forehead to the ground to pay respect to God, I couldn't help but inherit the humility my mother and father taught me. I always liked to dismiss it as the inheritance given by a professor to his daughter, but in truth, it was much deeper.

The call to prayer rang again, heavy in the air. With the legions of men and women around us, we stood facing the Ka'bah, even though we couldn't see it. As my mother had taught me to do as a young girl, we declared the intention of prayer with open hands at our ears. "Allahu Akbar," we each whispered to ourselves, starting our prayers and drawing our hands over our chests, right hand over left hand. "God is great." My mother taught me, as Muslims are taught around the world, to stare at one spot on the ground in front of me while praying, a focusing technique similar to what I learned in yoga.

Whispering under our breath, we followed the imam, or prayer leader, of the Sacred Mosque of Mecca as he recited the first chapter of the Qur'an—"Al-Fatihah" (The Opening). Akin to the Lord's Prayer for Christians, it was a recitation we used often.

In the name of God, Most Gracious, Most Merciful
Praise be to God, the Cherisher and Sustainer of the Worlds;
Most Gracious, Most Merciful;
Master of the Day of Judgment.

Thee do we worship,
And Thine aid we seek.
Show us the straight way,

The way of those on whom
Thou hast bestowed Thy Grace,
Those whose (portion)
Is not wrath,
And who go not astray.

My mother had taught me these simple verses in Arabic when I was a child in Morgantown, and I had committed them to memory, always seeking this elusive straight path. Now in Mecca I was again calling to God to guide me to the straight path. When I ended the recitation in concert with the gathered, our voices echoing in unison over our lines of congregation—"Ameen" (Let it be so)—the effect on me was profound. This awesome testimony to the union of people was magical and lovely. Having my family together as a part of that voice gave me a deep sense of unity with the strangers around me.

We continued with another chapter of the Qur'an. According to tradition, it can be any chapter. The imam recited one that I didn't know, and I followed his recitation with my eyes squarely focused on the ground in front of me. When I prayed alone, I had always picked the shortest chapter to recite, one of the first chapters I had memorized after "Al-Fatihah." This time, I recited the words of the imam.

"Allahu Akbar," we said, marking the end of this part of our prayer and moving to the next motion.

Together, we leaned over with our hands at our knees and repeated a proclamation of devotion three times. *Allahu Akbar* again. We stood straight. *Allahu Akbar.* We went to the ground, our feet beneath us, and pressed our foreheads to the ground. Three proclamations of another devotion. Along with the image of women wearing head scarves, this image of Muslims prostrating themselves in prayer is perhaps one of the most photographed images of Muslims because it reflects the total submission that the word *Islam* means. I knew I hadn't totally submitted to the faith, however, and the posture usually didn't feel right to me. *Allahu Akbar.*

We sat up. I turned my feet under me in the precise way that my mother had taught me. Pause. *Allahu Akbar.* Forehead to ground again. At this point I considered myself a Muslim yogi. Three proclamations again. *Allahu Akbar.* That was one *rak'ah* (unit of prayer) completed. There is a science to prayer in Islam. Each prayer has a specific number of *rak'at* (the plural of *rak'ah*). Some are *fard*, or mandatory. Some are Sunnah, or blessed but not obligatory. And some are *witr*, which means "odd number"; they represent an odd number of voluntary rak'at after the last prayer of the night. *Nafl* prayers are "extra" or optional actions. I became adept during a summer vacation when I stayed at my paternal grandmother's house and my youngest paternal aunt wrote the number of each prayer's rak'at into my childhood diary. The predawn prayer is two rak'at blessed but not obligatory followed by two mandatory rak'at. Yes, it is confusing. But somehow, like Masons with their ancient secret handshakes, we memorize these rites as part of being Muslim.

We returned to standing position. The imam led us through "Al-Fatihah" again.

"Ameen!" we all declared together, our unified sound merging with the wind. It was an exciting sound. The imam led us through another chapter. I didn't recognize it but followed silently.

At the end of the second rak'ah of this prayer, as in the final rak'ah of all prayers, we sat and said another required prayer. During this prayer we raised our index fingers while we recited words that included the *shahada*, or testimony of Islamic faith.

Ash-hadu anla ilaha ill-Allah (I bear witness to the fact that there is no deity but Allah)
Wa ash hadu anna Muhammadan 'abduhu wa rasuluh (and I bear witness that Muhammad is His servant and messenger)

From the corner of my eye I saw the index finger of the elderly woman next to me flutter upward while she made her proclamation of faith. My finger sat suspended in air too for a moment. It was quite profound that this woman and I were separated by culture, language, and economy, but we were both making a proclamation of faith at precisely the same moment. Bringing our fingers back to our laps, we quietly finished our Qur'anic recitation. We both turned to greet, first, the angel who sits on our right shoulder.

As-salaam alaykum (Peace be upon you)
wa rahmattullahi (and Allah's mercy)
wa barakatuh (and blessings).

I was always pleased to recognize the presence beside us of spiritual beings other than our bodily selves. Then we turned our heads to greet the angel who sits on our left shoulder with the same greeting. When my mother taught me this ritual, she instilled in me a consciousness that divine eyes are always watching us. "These angels are recording everything you do—good and bad," my mother told me.

We put our hands together in prayer in front of our faces. I prayed for peace of mind, called *sukoon* in Urdu, as I had done since my first prayers as a girl. To end prayer, my mother had taught me to *phoonk*, the Urdu concept for a breath of air meant as protection from harm. The word *phoonk* is a noun, but a dear friend of mine, Rachel Kessler, turned it into a verb not long after she learned about it. I had worked with Rachel in the Washington bureau of the *Wall Street Journal*, and she had supported me during the dark days of depression following the breakup of my marriage. She was Jewish American, but religion rarely entered into our discussions. We talked mostly about how to keep our souls intact and thriving as we tried to find love and happiness in the world. In Mecca I finished my prayer and sent a *phoonk* to the world. May we be protected and blessed.

LAWS OF MEN IN THE NAME OF GOD

There is no compulsion in religion.
"Al-Baqara" (The Cow),
Qur'an 2:256

MECCA—That night we returned to the Sacred Mosque for my mother's and Safiyyah's first entry. Sheikh Alshareef, our guide, led us through more stretching exercises. "C'mon, Mom! Stretch!" I yelled at my mother.

"Oh, Asra! I can't," she answered. She didn't consider it appropriate to be bending over, touching her nose to her knee, especially in this country that so controlled women's movement. I had spent my entire life doing stretching exercises like the kind we were doing to prepare ourselves for another circling of the Ka'bah. I grew up as an athlete with a real sense of ownership over my body. What I felt in societies such as this puritanical Muslim culture was that men had more control over women's bodies than the women themselves did. Men set rules and laws that defined women's reproductive rights, women's sexual rights, and, in a way that had proven deadly just the year before, women's right to free movement. To me, these restrictions not only defied internationally accepted standards for simple human rights and decency but also violated important tenets and traditions established at the time of the prophet Muhammad. So many rules are imposed upon us in puritanical societies as absolute laws of God when they are simply controls instituted by men. We have certain societal rules that bring order to our world. But it has long seemed to me that women, like men, should be able to live with rights to self-determination over their own bodies. To me, there is something fundamentally wrong about the way Muslim communities define themselves. It strikes me that women—and men too, because they are oppressed in a different way—should be bound by something greater than rules: the inspiration of living well. All around me in Mecca were reminders of the constraints imposed upon women in traditional Muslim society.

As we walked together to the Ka'bah, passing Nigerian mothers selling their wares without any male chaperones present, an elderly man and his thin wife passed by us. From their features, I recognized them as people living on the Afghan-Pakistani border in the Northwest Frontier province of Pakistan, where the most puritanical version of Islam defines society. The men there had just enacted rigid interpretations of sharia. The man wore a black turban, a trademark symbol of the Taliban. The woman looked like the first wife of the Taliban deputy ambassador I had interviewed in Islamabad after 9/11. She had the same thin face with a pointed nose and fair, weathered skin. There was something remarkable about this woman in Mecca, though: her face was bare, and she walked beside her husband, bumping into men as they wove through the crowds. When I went to a bustling city called Peshawar in Pakistan's Northwest Frontier, I wrapped my scarf around my face, only my eyes visible through a slit in the fabric, to follow the local tradition that required women to never show their faces in public. There was only one way this woman could publicly show her face to the sun, according to the men in her region, and that was by going on the hajj. Even the strictest of Islamic scholars, as well as my pocket-sized hajj rulebook, ruled that it was "forbidden" for women to do the hajj with face veils. Aisha, a wife of the prophet, had apparently lifted her veil when she did the hajj.

Two years before, I had ridden my motorcycle into my mother's ancestral village of Jaigahan, India, and walked about freely, wearing a scarf over my hair but no veil over my face, as all the other Muslim women had to do. "What is it like to feel the wind on your face?" a cousin had asked me.

I was stunned by her question. I had never fully appreciated the great freedom I enjoyed in being able to feel the sun on my cheeks. I had taken for granted the experience of having the wind and sun caress my face when I was outdoors. All my life I had enjoyed this luxury without fully valuing it. Before my jaunt through Peshawar, I had covered my face only three other times in my life. The first time was when I had visited my relatives in the university town of Aligarh in northern India. For a day I had worn the burka, the full gown and veil, as my cousins did every day. Though I realized that the veil couldn't deny the person I was within, I noticed that I retreated a bit socially because of the symbolic meaning the veil carried. I didn't feel I could truly be myself or fully express myself. After that I had covered my face again to shield my cheeks from the dust of the mostly barren Himalayan village where I had gone with my sister-cousins Lucy and Esther Ansari for the Buddhist holy pilgrimage of the Kalachakra. Lucy taught me how to turn any long scarf into a veil. Then, after I had returned to America, I veiled myself when I taught meditation to fourth-graders on Exploratory Day at Morgantown's North Elementary School—I even taught the eager boys how to turn themselves into little ninjas.

My mother was shocked at the feelings that coursed through her as we stepped into the Ka'bah. As a girl, my mother had known the strictest restrictions imposed on girls in traditional Muslim society. She had spent her earliest years in the one-road village of Jaigahan, where I would later ride my motorcycle and wander bare-faced. She would have never dared do the same when she was a woman there. Conservative Islam defined the community in that part of India, and her village was no different. Women received no more than the barest education, and men were mostly farmers. As a young girl, my mother ventured across a dirt path from her house to the threshold of the mosque only to deliver food prepared during the holy month of Ramadan by the women of the house—*channa*, or chickpeas, and *bhajiya*, appetizers made from chickpea flour—for the men gathered at the mosque to break their fast. Throughout her life her relationship to the mosque never changed. Even though my father helped start mosques, she never entered them. Like other women, she wasn't encouraged to pray in that space. That was why she was so profoundly affected by crossing the threshold of the Sacred Mosque of Mecca. It was only the fourth mosque she had ever entered in her almost sixty years. She said, "I didn't feel it was my right to go inside."

As we walked through the corridors I was impressed by the women resting on the floors of the Ka'bah between prayers. They were unescorted, and that was impressive in and of itself. Part of me wanted to be one of them. I wanted to sleep in the Ka'bah rather than in room 708 of the Mecca Sheraton. I wanted to know this level of surrender. Even my mother was impressed. "This is devotion," she said, looking at the women.

She remembered something Sheikh Alshareef had told us about the responsibilities that privilege carries: "We should be the first at the Ka'bah when we stay at the Sheraton." In fact, we weren't. We skipped the crowds to come only in the dark of the night lit by the thousands of bulbs in and around the Ka'bah. I held Safiyyah's hand and my mother's to guide them as Samir kept instructing them, "Close your eyes. Close your eyes. Close your eyes."

I put Safiyyah in place between two pillars, the Ka'bah in front of her. I released my hands from her shoulders. "Now, open your eyes!" Samir shouted. He felt a sense of privilege. Safiyyah's eyes fluttered open. There before her was the square building draped in black fabric. "It's supposed to be the house of Allah," I said, repeating what I'd read in my hajj material.

Safiyyah stared at the Ka'bah. "Did God really live there? Isn't the air his house?" I looked at her, startled. I appreciated her critical thinking. I felt the same way. I didn't believe an ethereal force such as God would be contained even symbolically in a man-made structure. The domain of the divine seemed more to be in the nature all around us. "Good point," I told Safiyyah. Samir contemplated the question. "But they can't build a house in the air," he said. "That's true too," I said.

What the children underscored for me was that, indeed, the Ka'bah is a human manifestation of the universal presence of the divine. Human beings put four walls on the house of Allah. Men and women attribute human traits and even gender to Allah when they declare His anger, His joy, His forgiveness. This has created a historical tension for those, from poets of yesterday to my mother and me, who resist human definitions of the spiritual and religious path, including a gender association with God. I don't consider God to be a He. To me, God is a force beyond gender and human emotion. To make such associations seems to limit the awesome force of creation. Many a Muslim mystic poet has written about the true Ka'bah being in our hearts. But what really meant so much to me was that we were having this conversation as a family—a man, two women, a girl, a boy, and a baby—in the most sacred place in our religion. This would not have been possible in so many mosques of the world where women and men, and boys and girls, are separated by man-made rules controlling movement and space.

We slipped into the courtyard and became a part of the pulse of humanity circumambulating the Ka'bah. Even though it was my second time going around the Ka'bah, I had to admit I really didn't understand what I was doing. Michael Wolfe had warned me that "confusion about the rites is universal." My mother was swimming in many emotions. She was awestruck that she was actually at this place she had heard about all her life. But more importantly, she was afraid the children would get crushed.

"Safiyyah! Safiyyah!" she yelled, as she saw a throng of pilgrims coming our way. Safiyyah cut the perfect figure for a petite gymnast, making her vulnerable to physical danger from this crowd.

We made it only once around the Ka'bah before my mother started feeling too claustrophobic. The crush at the point in the ritual of kissing the Ka'bah frightened her. "Let's leave!" my mother yelled. She wanted to go upstairs to the roof, where it wasn't so crowded. We shouldered our way out of the courtyard and breathed easier as we rode the escalators upstairs, winding our way to the top floor. This was remarkable because we were allowed to move freely, which is unheard of in so many mosques of the world that have designated sections for women, usually cramped and out of the way. The ordinary in Mecca is extraordinary in so much of the rest of the Muslim world.

On the roof I walked independently with my mother and Safiyyah, Shibli on my chest. My father and Samir were far ahead of us. I was struck by the devotion around me. As I studied the ritual scars on the cheeks of a black African woman, Shibli stirred restlessly, as he had done the night before, groping for milk. What to do? Where to go? I headed over to an empty expanse of the roof as my mother and Safiyyah continued walking. Eyeing a short bookcase piled with copies of the Qur'an, I nestled in front of it, in part to block the wind that was now rushing by us. How could I protect Shibli from this wind? I was confused. The wide fabric of my hijab was the perfect cover for my son. But if I dared remove it from my head, I would risk being arrested as a criminal.

I thought about the fourteen girls who had died trying to escape a fire that broke out at Intermediate School 31 in Saudi Arabia the year before, in March 2002. According to a report by Human Rights Watch, a New York–based human rights organization, eyewitnesses, including Saudi civil defense officers, reported that several members of the *mutawwa'in*—the religious police of the Committee for the Promotion of Virtue and the Prevention of Vice—interfered with rescue efforts because the fleeing students, age twelve to fifteen, were not wearing the *nikab*, the long black cloaks with head and face coverings that were obligatory public attire for girls and women. Some even beat girls who were evacuating the school without proper dress. A Saudi journalist told Human Rights Watch that the mutawwa'in at the scene also turned away parents and other residents who came to assist. The *Arab News*, a newspaper in Jeddah, concluded that the mutawwa'in were at the school's main gate and "intentionally obstructed the efforts to evacuate the girls. This resulted in the increased number of casualties." The religious police reportedly also tried to block the entry of civil defense officers into the building. "We told them that the situation was dangerous, and it was not the time to discuss religious issues, but they refused and started shouting at us," *Arab News* quoted civil defense officers as saying.

I had heard about this case, and I considered it tragic testimony to dogma overriding common sense—an apt description of my own situation at that moment on the roof of the Sacred Mosque, with my son getting cold in the night wind. A woman rubbed her arms with her hands as if to gesture that I needed to cover Shibli.

"I know. I know," I said to the wind in frustration.

But I dared not remove my hijab to warm Shibli. *What am I supposed to do?* I thought to myself. With the Ka'bah below us in the courtyard, I nestled Shibli under my hijab and let him find the warmth of my breasts. He nuzzled against me, nursing. I felt calm for that short time. After Shibli dozed off, I stood up to try to find my family. I circled half the Ka'bah and found myself again at the place where Hajar had run. Settling with my back against a wall that marked the hill of Safa, I stared at the paisleys on the scarves of women in the sea of worshipers around me. I wondered about the women who drifted by me, all so anonymous but all united by one mission, to move between these two hills. I knew that I was very different in many ways from the women I was watching. We looked different. We talked differently. But we were the same in that we were enjoying some of the greatest freedom of movement allowed throughout the Muslim world while still facing immense constraints. I stayed like that for hours, absorbing and appreciating the expression of movement around me.

So often it seemed the law cracked down on women trying to exercise freedom of movement. For instance, women weren't allowed to drive in Saudi Arabia. On November 7, 1990, during the buildup of U.S. troops in Saudi Arabia for the Gulf War, forty-seven women from the Saudi intelligentsia challenged the unofficial ban on women driving by taking their husbands' and brothers' cars out for a drive in Riyadh, the capital. Mostly Western-educated, they had driver's licenses from other countries. The Saudi religious elite struck back with a fatwa against women driving, and many of the women were detained, lost their jobs, or had their passports revoked for two years. The king made the fatwa the law of the land.

I considered these rules some of the most imposing restrictions on a woman's right to self-determination. I knew the power of unfettered mobility. The year before the women's protest drive in Saudi Arabia, I had scrambled through my Chicago apartment to find my driver's license, lost in one of my pants pockets. That was the summer when United Airlines flight 232 crashed in a cornfield outside Sioux City, Iowa, killing 112 people. My editor at the *Wall Street Journal* had dispatched me, the cub airline reporter, to report from the scene, my first reporting assignment in the field. I had to fly to Omaha, Nebraska, to rent a car and drive Interstate 29 north to Sioux City. Breathing a deep sigh of relief when I found my driver's license, I bolted out the door to go do my job. My driver's license empowered me as a professional journalist and as a woman.

I could never have asserted myself in Saudi Arabia as I'd done in America. By law, Saudi women aren't allowed to go out in public or out-

side the country without a male chaperone. To me, this was unconscionable in a religion that sprang from the courage of a woman who lived
alone in Mecca. Hajar certainly had no male chaperone. The U.S. State
Department on its website warns travelers that a married woman residing
with a Saudi husband can't leave the country or have her children leave
Saudi Arabia without her husband's permission, even if the woman or her
children are U.S. citizens. The husband, as the family's sponsor, is the
only one who can request an exit visa. In most of the country women
can't go out without being covered from head to toe in black. The State
Department even advises women not to wear pants. As pilgrims, we were
excused from wearing the fully veiled nikab but were advised not to walk
in public without a male escort.

The night slipped into the early morning, and Shibli awakened with
the azan, the call for prayer. It was a magical moment as the nighttime
flock of ababeel gave way to the morning's birds. A few dark clouds hung
in the air. A row of Indonesian women in lacy white hijab stood nearby.
They caught sight of Shibli and smiled. A woman prayed from her
wheelchair. With Shibli awake in my arms, I couldn't join the prayers.
Men and women were standing shoulder to shoulder, leaving no place to
lay Shibli down. I stood alone as this sea of Muslims stood, bent, and pros-
trated in unison. After the prayers had ended, a Turkish woman living in
Germany put a shimmering brown scarf in front of me to pray upon. I
looked at Shibli, wondering how I would pray with him. "I will hold him,"
the Turkish woman offered.

I hesitated. What if she ran away with him? I yielded to faith and put
Shibli carefully in her arms. I prayed, greeting the angels on both my
shoulders with the conclusion of my prayer. I turned to find Shibli nestled
in the Turkish woman's arms. He was safe. I eased him back into my arms.
By then, my family had found me. As we left the Turkish woman purred
with a sound that Shibli appreciated. I was struck by the universal sound
of love. A man from Indonesia stopped to play with Shibli. "As-salaam
alaykum!" he said loudly and warmly. Shibli looked warmly into his eyes.
The man and his wife missed their son, whom they had left at home to do
the pilgrimage. She played with Shibli too. I was struck by the friendliness
with which pilgrims, even male pilgrims, greeted us. There were few formalities and barriers. As I left I had a warm feeling of acceptance and
freedom. My father and Samir headed back downstairs to retrieve our sandals. My mother, Safiyyah, Shibli, and I boldly did the unthinkable. We
headed home alone to room 708 of the Sheraton without our requisite
chaperone.

THE PROPHET AND HIS FEMALE ANCHORS

MECCA—When I looked at the forgotten history of Islam, I saw the legacy of women who had to stand very much alone in the world, starting with the mother of the prophet of Islam.

The prophet was born in the year 570 in Mecca and named Muhammad ibn Abdullah, or "Muhammad, son of Abdullah." His parents had married not long before, but his father didn't see his son into the world. A merchant, he died just days before his son's birth on a trading excursion. Interestingly, Muhammad's father's name, Abdullah, meant "servant of Allah," after the pagan deity with that name at that time in pre-Islamic history. Although the history books don't cast her this way, his mother, Amina, entered motherhood as a single mother. I had always heard this story, but it took on special meaning for me after I took my own son through the city of the prophet's birth. I knew well the fears and dashed hopes of a mother bringing a child into the world alone, no matter what the circumstances. At a time when I was looking for strength from women in my religion, I found a strong woman right at the birth of Islamic history, namely, Muhammad's mother.

For reasons that aren't explained, Muhammad's mother died when he was six years old. Historians don't speak much about her life or her influence on her son. The Qur'an doesn't mention her by name. But having seen through my own eyes the immense influence of a mother on a child in the early years, I can't help but think that she must have had a profound influence on the man he grew up to be. Much as my son went to my father's home while I worked, Muhammad went to his maternal grandfather's home after his mother's death. Sadly, his grandfather died three years later, and Muhammad, at about the age of my nephew, Samir, was then sent to live with his paternal uncle, a struggling trader. The family, the Hashmi, were merchants; though not rich, they belonged to the tribe, the Quraysh, that had sprung from the people who had found Hajar and her son Ishmael in the desert generations earlier. They were now the leading trading tribe of Mecca.

The young boy traveled with his uncle on caravans throughout the Middle East. When his uncle's tottering business finally failed a few years later, Muhammad returned to Mecca, where he became a lowly goat herder. Sometime later, Khadijah, a prosperous woman who led a trading empire, was looking for a man to lead her caravan to Syria. I always thought of her as Leona Helmsley without the mean streak or the IRS problems. Muhammad's uncle got him the job.

Mecca was a rich city, but this wealth had a dark side. There was a great divide in Mecca between the rich and the poor, an underclass of slaves and hirelings, and rigid class barriers. A loose council of elders from the well-to-do families ruled the city and tried to enforce law and order, but they spent a lot of time fighting among themselves. In that culture, Muhammad came to be known as an honest broker.

Islamic historians say Khadijah respected the virtue of her young employee and proposed marriage to him. Muhammad was twenty-five and single. Khadijah was forty and a widow. She had no fear about making the first move, however, and he had no fear of a working woman who was more successful than himself. On the most intimate level, Khadijah chose her lover. She was a woman who had ownership over her heart and body. That ancient proposal is the first example we have in the Muslim world of the right of women to self-determination in matters of love. With his acceptance of the proposal, Muhammad blessed this notion. In contrast, I've read Pakistani newspapers filled with accounts of the horrors of fathers, uncles, and brothers throwing acid on the faces of women in their families who acted as independently as Khadijah. According to the Human Rights Commission of Pakistan, in 2003 nonprofit groups in one of Pakistan's largest cities, Rawalpindi, said they recorded up to five thousand cases of women being burned alive, often for claiming their rights, by their husbands or in-laws over the previous five years.

I truly came to admire Khadijah when I read *Nine Parts of Desire* by Geraldine Brooks, my former colleague at the *Wall Street Journal*. Geraldine cites a saying around the time of the prophet Muhammad that women have responsibility for nine parts of desire and men responsibility for only one part. She speaks of Khadijah as the prophet Muhammad's "boss." I had never thought of their relationship that way before. Muhammad and Khadijah would have at least six children, of whom only four daughters survived. He married no other woman while she was alive, though he did take multiple wives later.

Khadijah was a key figure in the prophet Muhammad's success. Mecca was a decadent and unruly place, but Muhammad had a reflective personality and Khadijah gave him the space to express it. Beginning in his early thirties, he often ventured out of town to meditate in a cave atop a small mountain called Mount Hira, or Jabal al-Noor, "Mountain of Light." At the age of forty, he experienced a transformative moment. On the seventeenth night of the month of Ramadan in the year 610, he ascended to his cave to pray, fast, and meditate, as he had grown accustomed to doing every year during that month. He was worried that year about the spiritual emptiness that had crept into the business culture of his tribe, the Quraysh. While worshiping the high god named al-Lah, meaning "the God," they threw their wealth at the worship of many gods. Pilgrims swept through Mecca every year to worship trees, stones, and wells inhabited by powerful spirits. Consistent with the roots of many religions, most of those spirits had one thing in common: they were female.

By that time in history, monotheism was alive and robust. Judaism and Christianity had been around for centuries. Jews and Christians lived freely in Mecca, doing business with the people of the Quraysh tribe and the rest of present-day Saudi Arabia, and they often mocked the Quraysh for not having a prophet or divinely ordained path.

That night Muhammad awakened to hear words spilling from his mouth. Those words were the first chapter of the book that would become the Qur'an. It is said that the angel Jibril appeared to Muhammad and told him to preach to the people about Allah, the one and only true God. Gabriel told him that he was chosen to be the last prophet of God. Fear gripped Muhammad. He responded with the age-old remedy: flight.

He ran home and hid under his bed sheets. The next day Khadijah—not afraid—took her husband to meet her aged, blind cousin, Waraqa ibn Nawfal, a Christian man who knew how to read and write. Waraqa told him that the angel who had visited him was the same one who had visited Moses and Mary, mother of Jesus. Waraqa told Muhammad he would be a prophet but would be opposed by many. It took a woman—Khadijah—to act clearly, bravely, and wisely enough to encourage Muhammad to venture onto the world stage as a prophet.

And so Islam, its name meaning "surrender," was born. Its full meaning suggests submitting oneself totally to God. Muhammad's chief message to his people was that "there is no other God but Allah." He taught a monotheism that challenged the polytheism of the time. Khadijah became the world's first convert to Islam. With her wealth, she provided Muhammad with the material comfort that allowed him to focus on his spiritual practice, meditation, and preaching.

In one respect Khadijah was not unusual: working women were an integral part of the earliest history of Islam. I was stunned when I read about their contributions to early Muslim history. They were fully engaged in society. A Muslim woman named Samra bint Nahik al-Asadiyya cracked a mean whip that she literally kept with her as she carried out her job in the market as a *muhtasib*, a market inspector who made sure people had ethical business dealings. The divorced aunt of a man by the name of Jabir ibn Abdullah left her house to get some of her date palms harvested and sold. Someone tried to stop her, claiming she wasn't allowed out during the period of *iddah*, the time a woman is supposed to wait to remarry after being widowed or divorced. She went to the prophet for a verdict. He told her: "You go out and get the dates harvested (and sold) so that you may be able to do some other good work." Muslim women supported their households. The wife of a famous scholar, Abdullah ibn Mas'ud, told the prophet that the family's only income came from her handicraft work. "I am a woman engaged in handicraft (on which we survive), but my husband and son have nothing," she said, sharing with him her life as a working woman. Muhammad didn't admonish her.

After Muhammad recited the verses on women and men living modestly, Umar, a leading companion of the prophet's at the time, saw a woman named Saudah leaving her home. He stopped her and admonished her for not covering her hair. She continued anyway but told the prophet about the scolding when she returned home. He listened and responded, "Undoubtedly, Allah has permitted you to go out to fulfill your needs." The prophet even encouraged women to work. Qilah, a *sahabiah*, or female companion of the prophet's, came to Muhammad to ask various questions on business practice and said, "I am a woman who purchases goods and sells." His acceptance of her was an affirmation of women working outside the home.

Even after the prophet's death, there were reports of women who made their livelihood in business. When Umar served as caliph, or leader of the Muslim community, a woman named Asma' bint Makhzumah used to sell perfume imported from Yemen. Another companion of the prophet's once praised the negotiating skills of a woman who bought a big fish at a reasonable price.

On our third day in Saudi Arabia, my family and I stood in the busy street outside the Ka'bah and tried to find the place where Khadijah lived with the prophet. "Khadijah lived here somewhere," my father said, pointing toward rows of narrow storefronts in front of which a flurry of cars swept by, beeping loudly. With those historical images in our minds, we headed out to the mountain to pay our respects to the place where historians say Muhammad received the revelations that became the Qur'an. On our way, we passed a storefront for Daewoo Motors. At the base of the mountain, next to a Toyota and Mazda dealership, the driver stopped beside signs for GTX and Castrol motor oils. We craned our necks to look in awe at the steep climb that some pilgrims were making to the Cave of Hira, where the prophet used to meditate. As we pulled away the driver told us that Khadijah used to walk the many miles from Mecca to the cave just to give her husband his meals.

My father listened intently to the driver's story. He was so moved by Khadijah's commitment to the prophet that he started to weep. Safiyyah, Samir, and I exchanged curious and worried looks with each other.

Today Khadijah is remembered at a grave that sits behind a boundary wall in a cemetery in Mecca called Jannat al-Mualla, or "the Cemetery of the Exalted." It is a magnet to the devout, especially those who respect the power of Khadijah's spirit. It is near the Mosque of the Jinn, where we were told a group of jinn, or "spirits," came to the prophet Muhammad and listened to him recite a portion of the Qur'an.

We had to be careful how we expressed our respects at Khadijah's grave, which we discovered to be merely an unmarked patch of dirt with small rocks piled on it. On the way to the cemetery we had been given an innocent-looking brochure with a subtle message: "May Allah Accept Your Hajj." Inside, the brochure spelled out the rules clearly: there were to be no acts of devotion at any graves. The Saudis banned paying homage at graves, and, indeed, the Wahhabi ideology had emerged in part to crush homage given to Sufi saints, a practice that was often carried out at their graves.

"Bebe! Bebe!" a woman with the air of nobility shouted to Shibli as we stood in the hot afternoon sun at Khadijah's grave. She was a pilgrim from Nigeria. We were both there to pay respects to the great lady of Islam. Before setting out that day, the Nigerian pilgrim had wrapped a batik scarf around her head in a covering that seemed almost royal. She sat on a bench and reached out, enfolding Shibli's delicate fingers into her dark hands. As she stood up she towered above us. She took Shibli into her arms and nuzzled him against her bosom. Laying him back into my arms, she ran to catch her bus. As she took her seat I waved good-bye with Shibli in my arms. As she leaned out of the window of her bus I told my new friend that maybe I would visit her one day in Nigeria. We had had an ordinary encounter that lasted only a few minutes, but meeting her was meaningful to me. In embracing Shibli and me, she epitomized, like Khadijah and Mary and Muhammad and Jesus, the true spirit of religion: love and kindness without preconceptions and judgments. It seems to me that we need more of this spirit in the world.

"I'll be expecting you," she said.

I breathed the air deeply at Khadijah's grave, trying to absorb the great spirit of the woman who had been buried there. My mother, who hadn't

joined us for this pilgrimage to the grave of Khadijah but instead was resting from the night before at the Ka'bah, didn't need to breathe in this
powerful woman's spirit. She lived it.

THE COURAGE OF MY MOTHER

MECCA—It saddens me to realize that Khadijah could never have lived in
modern-day Saudi Arabia. The Saudi government doesn't allow women
to run businesses in their own names. When I opened a local paper, the
Riyadh Daily, I read, to my horror, that the trend of Saudi women working
as babysitters was so noteworthy it made headlines: "Saudi Women as
Babysitters Getting Common."

The article said Saudi women wanted to make money caring for children inside their homes. A teacher protested the trend: "How can a
mother allow her daughter to work with people she does not know and
how can the father or husband allow this?" Her rhetorical question had its
answer in history. They can. Khadijah would never have met the man she
nurtured into a prophet if she hadn't worked with people she *did not know*.

Like the ban on women business owners and the discouragement of
working women, so many modern Islamic traditions that are imposed
upon women contradict the practices that existed in the seventh century
when Islam first emerged.

My mother was born into a traditional Muslim family where women
didn't work for others. They worked only in the home for their husbands,
children, and families. Her family observed a strict gender segregation, or
purdah. Literally "curtain," purdah effectively separates men and women
into "domestic harems." In my mother's family this strict gender segregation stayed in place even at times of death. When my mother was about
two years old, her father, Ali Ansari, fell suddenly ill. Her mother, Zohra
Ansari, kept leaving the room when men visited her ailing husband. Even
at that critical time, she didn't dare cross the hudud, the boundaries. As a
result, my maternal grandmother was in another room when her husband
took his last breath. Years later she told a relative that she lived with the
regret of not having been with her husband when he died. As my mother
grew up she saw women scamper into a back room whenever men unrelated to them arrived for a visit. They rarely debated politics, business, or
religion with men to whom they weren't related, and hardly even when
they were. My mother inherited a deep sense of hudud. From her teen

years, she covered herself with the traditional black burka that covers the face and body in a gown.

But my mother was always a rebel. As a young woman, she irked an aunt when a servant revealed that my mother and a cousin had removed their burka at their women's college, Nirmala Niketan. (They did it to avoid social embarrassment among their less traditional classmates.) She wasn't allowed to complete college and was soon married off instead to my father in an arranged marriage. When she disembarked onto the train platform in the city of Hyderabad, she got a strong signal about the turn her life had taken. Her new mother-in-law, a more liberal Muslim woman than her own mother, took off her burka. Stunned, my mother never wore the burka again except when she visited her family.

Transformed dramatically in the first days of marriage, my mother looked as fragile as a china doll in her newlywed photos, but she had a strong will that defined the rest of her marriage. She clashed quickly and regularly with the traditional expectations of my father's mother. And yet my mother was dependent on her mother-in-law and the rest of my father's family for her well-being. When my father left for his first jaunt to America, she vowed she would never put herself in such a position of dependency ever again. For my mother, America held out the same promise it had for countless other immigrants: intellectual freedom, economic independence, and personal liberation. At the huge emotional cost of leaving her children behind until they could afford to reunite us, she left to join my father in America with these dreams tucked secretly into her belongings.

If the traditions of Saudi society had applied to my mother in America in the 1960s, it probably would have been many more years before I was reunited with my parents in America. To raise money for airplane tickets for my brother and me, my mother ran a babysitting service out of their house in Piscataway, New Jersey, caring for children named Eda, Laura, and Chris. Laura called my mother "her mother in the red house." Another girl sent my mother into a panic one day when she put Vicks on her face. Fathers sometimes dropped their children off when my father wasn't home; my father was neither threatened nor offended by that. My mother retained her sense of modesty, but even such interactions were a huge departure from her upbringing in India.

After my brother and I arrived, my mother continued to babysit to make money. Business was in her blood. An older brother, Anwar Ansari, had started a wholesale business in India manufacturing and exporting clothes, like the embroidered kurta (or tunic) popular with hippies in those days. Though, as far as she knew, few woman in her family had ever worked outside the home, my mother, unfettered in America, sold these shirts. With the family and boxes of kurta piled into our green Rambler station wagon, my parents carted folding tables to the Rutgers student union in New Brunswick and a flea market called Great Eastern. My brother and I enjoyed the thrill of our weekend outings. My mother mingled equally and easily with women and men, and my religiously devout Muslim father had no conflicts of faith over my mother's work. After a few years my mother helped her brother open a showroom, India Village Industries, on Broadway in New York City. There she worked all day with hard-bitten managers from India, negotiating with retailers and inspecting supplies.

When we moved to West Virginia in 1975, she got a job as a lab assistant in the home economics lab of West Virginia University, where I would visit her after school. A surly boss drove my mother from that job, however, and she returned to the home, where she guided me as always in my academic and religious life while she completed her BS degree at West Virginia University.

In the summer of 1981, with my father's encouragement, my mother stepped into a new role. First, she waved good-bye to me as I sat in a jet on the tarmac in Pittsburgh, about to fly to Oklahoma for a monthlong science camp sponsored by the National Science Foundation. We were so close: she wept as she saw me off, and inside the plane, at sixteen years old, I wept as I left my mother for only the second time since we'd been reunited in America. (The first time I had visited relatives.) When I returned, my mother was caught up in a flurry of activity: she was opening a boutique on busy Walnut Street in downtown Morgantown, where she would sell more of the kinds of clothes made in India that she had sold in New Jersey.

My father supported her fully. He worked beside her late into the night designing a sign for the storefront and advertisements he hung all around town. I had pushed my mother to spend so many weekends at the flea market that she named her boutique Ain's International after me, using the last syllable of my middle name, Quratulain, which means "coolness of the eye" (*ain* means "eye"). Starting with a deep-throated vowel sound, it proved to be a tricky name for Americans, but my mother was fine hearing my name pronounced like the English "Anne." She reminded me that her brother used to call me Annie. Late that summer, just as university students returned to campus for the fall semester, my mother opened the doors on a beautiful store with rugged wood paneling and racks of beautiful flowing cotton and silk skirts from India. After school I'd spend the late afternoon there.

I felt proud of my mother as she asserted herself as an independent businesswoman. Watching her realize her dream in the public world encouraged me to strive to fulfill my own dreams. That year I was editor of my high school newspaper, *The Red and Blue Journal*, and I often laid out pages in the back of my mother's boutique. She never complained about the overpowering scent of rubber cement glue that wafted into the store. Over the years, from Bombay to Broadway, I watched my mother haggle with wholesalers with charm, a stiff spine, and, perhaps most important of all, her feet ready to walk out the door if they didn't compromise. This was bold work for her to do coming from a family where women weren't so openly engaged with men.

In her boutique my mother was like a second mother to many West Virginia University students away from home, and she was a friend to many of the Morgantown locals, some of them ex-hippies who believed in tolerance, peace, and kindness. What I appreciated about my mother was that during slow moments she would carefully read books she had pulled from my bookcases, such as the writings of immigrant women from India like Bharati Mukherjee and feminist writers from the Muslim world such as Egyptian Nawal El Saadawi. She had an open, literary mind. One day, as a young adult, I turned to my mother and said, "You are like a modern-day Khadijah." She dismissed the idea immediately, returning to the task at hand: making that night's dinner.

On the hajj I came to fully realize that my mother is a wise and courageous woman in many ways. She is a woman who will hunch her back rather than sit up straight. I am a daughter who will scold her to straighten her back. I realized, however, that we can learn from the people who show courage without demanding perfection from them. We often have a guide when we embark on a journey, and she was mine. We need to seek out courageous people to be our guides, and we need to be courageous ourselves to fully learn from them. For me, my mother was a simple but brave guide, especially at moments such as her Kentucky Fried Chicken adventure.

On our second night in Mecca, my nephew, Samir, was hungry. "Let's go to Kentucky Fried Chicken," he said to my mother in the hotel room. Back at home my mother wouldn't have hesitated, but here she calculated the risks carefully. My father, her authorized male chaperone, wasn't around. Since he was under the age of puberty, Samir wouldn't qualify for the role. My mother would be violating the Saudi law that requires an authorized male chaperone for all women venturing out in public.

I chafed at the bridle that constrained women's free movement in this Muslim country. It made me angry, and the contradictions in how this law was enforced baffled me. All around us on the streets of Mecca, directly outside the Ka'bah, dark-skinned women peddlers from Nigeria sat with babies swaddled to their backs and goods spread on blankets in front of them. With no men around as chaperones, they were unescorted, independent women earning a living. Nobody disturbed them. But other women had to worry. Our first night in Mecca I had gone to a pharmacy down the street from the Sheraton with my father. As he dawdled, I felt it was patronizing and paternalistic that I had to risk arrest if I went to the next store without him. I went anyway, but made sure my father joined me quickly.

In the Mecca Sheraton, my mother wasn't about to let her grandson stay hungry. She gathered up her resources and pulled her hijab over her head. "Okay, let's go," she said.

As they crossed the street in their stealthy KFC run, Samir looked both ways for any religious police ready to swoop down on them. He urged my mother to move more quickly. "Hurry, Dadi! You might get arrested!" In front of the counter at KFC, Samir felt scared. "Are people going to say, 'You can't come here'?"

They ordered carryout.

THE CONTRADICTION OF PUBLIC AND PRIVATE

No matter how repressive the state became, no matter how intimidated and frightened we were, like Lolita we tried to escape and to create our own little pockets of freedom. . . . An absurd fictionality ruled our lives.
Azar Nafisi,
Reading Lolita in Tehran (2003)

MECCA—It was becoming clear to me that many Saudi women lived a different private life away from detection.

The rear of the Sheraton opened into a multistory shopping mall complex with an escalator running through it. It looked like any shopping mall in the West, but there was a strange mix of messages in this Saudi shopping mall. "Remember Elvis," screamed a T-shirt, snug on a busty female mannequin. The silver lettering of the words encircled a broken heart. A woman from Indonesia drifted by with a pink gift wrapping bow on her head, to help her group spot her from a distance. She had a hand towel over the bridge of her nose, covering the lower half of her face, with only her eyes peering out. In one shop pairs of high-heeled shoes labeled "Made in Italy" filled a shelf. A woman in a black burka wandered by a store window filled with sequined and denim purses.

We rode the escalator upstairs to a store with racks of blue jeans. I went through the racks and found a pastel green sleeveless nightie with a netting top. "Always New Fushon in Best Quality," the tag read with a unique spelling for "fashion." It was floor-length with a trim of green fur and a slit down the front. Safiyyah went through the racks too, eyeing the lingerie. She pushed away a red-and-black, hip-length nightie with a feather fringe. She stopped at a sheer peach gown with a matching sheer robe. "I think this might be an option for you," she deadpanned.

Like her, I was shocked at what we were seeing sold on the open market. I stopped in my tracks in front of a tight, sky blue shirt with a plunging neckline. The only thing halal (based on Saudi law) about it was its long sleeves. Many of the words emblazoned on its front held double meaning not only for me but also for this strict Muslim society in which I had found myself.

MEMORY

ECSTASY

TYRANNY

HYPOCRISY

UNITY

NOTORIETY

NO TIME TO THINK

Exhausted, we took refuge on the yellow-and-blue-striped sofas that lined the seventh-floor lounge of the Mecca Sheraton. Canadian pop star Avril Lavigne smiled out at us from a full-page feature on the MTV Asia Awards in the *Riyadh Daily*. "Look! Short sleeves," I told Safiyyah, gesturing to the singer's bare arms. The image was shocking to me because we didn't dare bare our arms in public in Saudi Arabia. It was illegal. I thought even the newspapers would have to comply with the rule, but clearly the mass media here wasn't held to the same standards that ruled the lives of ordinary people. The newspaper's TV guide listed *The Oprah Winfrey Show*, with all of its mini-documentaries on American societal drama, on Star World. In a country where homosexuality was illegal and punishable by death, *Will and Grace*, the sitcom about a gay man and his female roommate, could be seen on the Comedy Channel. Our first night in Mecca, my mother and Safiyyah spent the night watching a Saudi soap opera whose drama rivaled that of any American soap opera.

Saudi Arabian society seems to be defined by these contradictions. For women particularly, but not exclusively, the restrictions and repression breed not always compliance but rather conflict and dissonance. I know this because I lived this way myself for a decade, from my late teens into my late twenties. I lived a double life, secretly satisfying my curiosities about men while lying to my parents because I knew that I was crossing boundaries that weren't supposed to be crossed. I couldn't live with the lies, deceit, and hypocrisy after my marriage fell apart, when I realized that we aren't meant to suffer so deeply just to deny our true selves and realize societal, parental, and external expectations for ourselves. I decided then that I wasn't going to live with contradictions in my own life.

In any society governed by oppression and rules that don't make sense, there will be rebellion, even if it's expressed privately. To express such rebellion publicly is to me the sign of a mature individual and a mature society. From my experience, public disclosure allows for healthier expression and resolution.

RELIGION, SEX, AND SEGREGATION

MECCA—One afternoon my father and I ventured through the streets of Mecca to find a first cousin of mine who was on the pilgrimage with her husband.

After much searching, we found a dingy, narrow stairwell in the lodging house where they were staying. As the fat-cat pilgrims from America, we had pristine and immaculate accommodations at the Sheraton. They stayed in the accommodations that the government of India had secured in Mecca for its pilgrims. Most countries had similar accommodations, with buildings designated by nationality. I crept inside, gingerly. Oil streaked the walls near a gas stove that pilgrims used to cook meals. Here there were no Sheraton buffets. My cousin stood in front of the fire, frying an omelet. "As-salaam alaykum!" she cried out, pleased to see us.

"Walaikum as salaam!" we answered.

She led us into a small room with half a dozen cots lined up at different angles to each other. To my shock, men and women, unrelated to each other, were sleeping in the same room. Her husband sat at the edge of one bed, while a woman pilgrim lounged on another cot. The other men were in various stages of getting up from their sleep. Everyone sat up to welcome us into their space. My cousin introduced the pilgrims to us. The men were certainly not her mahram, but somehow, under the circumstances of the pilgrimage, it was halal for them to sleep in a mixed-gender room. Everybody seemed comfortable with the arrangement, and the women talked freely with their faces uncovered and only their hair loosely covered with scarves. Seeing my cousin accept this kind of living arrangement in Mecca was even more meaningful to me because she and I shared the same cultural programming. She was my ancestry. She would never have lived in the same room as male strangers in Bombay, but somehow she transcended all of her programming to do so in Mecca.

My experiences with my mixed-gender tour group, my forays into the Sacred Mosque where men and women prayed together, and my look at my relatives' pilgrimage accommodations revealed to me the inherent contradictions in Muslim society. Men and women mingled comfortably in Mecca. How could men and women be equal and interact without this burden of sin in Mecca but not elsewhere? This question had a profound impact on me. Places like Saudi Arabia strictly segregate men and women, but in the West Saudi men and women freely circulate in mixed-gender situations. Even where Muslim women are allowed to work in the West, the attitudes of segregation still prevail in traditional communities; even at dinner parties in Morgantown, working women accept sitting separately from men.

Religious dogma reaches into the most intimate corners of our lives. Most religions have repressive rules about sex except in marriage. Muslim leaders translated these attitudes into rules about women's behavior in public, and now sexuality has become repressed in Muslim societies. This association of sin with sexuality makes women feel a shame about their sexuality and sensuality that is yet another factor in the inability of Muslim men and women to intermingle innocently.

I believe this repression creates fears that are manifested in dysfunctional ways. A renowned twentieth-century Syrian poet, Nizzar Kabbani, noted the sexual double standards that emerged with the idea that women must be protected while men are free to wander. From the perspective of a woman, he wrote, "My brother returns from the whorehouse proud and strutting like a cock. Praise be to God who created him out of light and us out of vile cinders and blessed be He who wipes away his sins and does not wipe away ours."

I saw too that a hypersexuality emerges with the strict segregation of men and women. I saw that in India, where men and women don't mix as freely as they do in the West. Sexually repressed young men went out of their way to grope any woman walking through Hazrat Ganj, a bustling business corridor in the city of Lucknow. India has even established laws in recent years to prosecute men for "Eve teasing," its slang for sexual harassment. Traditionalists claim that strict segregation of men and women protects women from this kind of harassment. But I consider segregation a cop-out and not an effective response to the challenge of creating societies with healthy gender dynamics.

If men and women can pray together in the Sacred Mosque of Mecca, why can't they do so elsewhere? Surely, it seemed, if gender equality was good enough for Mecca it was good enough for far-flung places like my hometown of Morgantown.

A ROAD AND RECLAIMED HISTORY

ON THE ROAD FROM MECCA TO MEDINA—The prophet's early preaching at Mecca met with limited success: during the first three years he gained only thirty converts, a modest beginning certainly. His own tribe of the Quraysh persecuted him and his followers because of the revolution he was fomenting. Meccans attacked Muhammad through a fierce smear campaign designed to tarnish his reputation, and Muslims met in secret for fear of torture or even death. The followers of Muhammad, known as his *sahaba,* or male and female companions, practiced their religion for thirteen difficult years amid constant physical and mental torment. They couldn't build houses of prayer, and they didn't dare to pray at the Ka'bah. When they did, they were immediately punished. To the people of Mecca, Muhammad represented a threat to the status quo. He was preaching new rights for women, including their right to inheritance, political assembly, marital respect, and divorce. He was delivering sermons rejecting the polytheism practiced at the time and leading followers to believe in a monotheistic religion of faith in one God.

But Muhammad persisted, and gradually his following grew. Indeed, it grew so much that the powers-that-be in Mecca started to get alarmed. They sent a delegation to Muhammad and promised him great riches if he would just stop preaching. But Muhammad wouldn't be bought off. "Woe unto you idolaters," he told the city's officials.

The authorities realized they were going to have to take more definite action if Muhammad was to be stopped.

In 622, the Quraysh decided to move against the Muslims with the force of arms. Muhammad and his followers fled to Medina, about 277 miles to the north, where the prophet had been invited by city officials to mediate a civil dispute. The prophet and his close companion Abu Bakr climbed a mountainside outside Mecca to a cave where he hid from his enemies. In Islamic history it is recounted as a moment of divine protection. A spider is said to have spun its web quickly over the entryway, fooling the enemy soldiers into thinking the prophet Muhammad could not have crossed through there.

For me, the hajj was much deeper than religion and faith. It was like living history. I had never understood the concept of the Islamic calendar until I set off on the journey from which it started. Muhammad's migration to Medina fourteen centuries ago marks the first year of the Islamic calendar. The Islamic Society of North America advertised its 2003 hajj itinerary as "Haj 1423 A.H. Program"; A.H., or "after Hijrah," means after the emigration of the prophet Muhammad from Mecca to Medina. The year represents an important turning point in the history of Islam: Muhammad left his hometown defeated for a new beginning.

Like those first Muslims, we headed north to Medina, pulling away from the Mecca Sheraton as the crowds flooded to the Ka'bah for the early afternoon prayer. As our Mercedes Benz bus wound along the road to Medina, Tom and Jerry waved to me: the Saturday morning cartoon characters fluttered on a blanket hanging from a balcony. I dozed off and awakened to see the splendor that is my son. He was curled up in my arms, playing gently with his own lip. The name of the explorer Marco Polo stretched along the side of our bus, and I thought about the historical journey on which Shibli and I were embarking. We were following in the footsteps of the prophet when he went on the *hijrah*, or "migration," from Mecca to Medina.

At a rest stop a Chevrolet Suburban with bedrolls and canvas bags tied to the roof pulled in beside us, and a group of African pilgrims tumbled out. Back on the road, an open pickup truck glided by our bus, with camels inside. "Camels!" I whispered to Shibli. Nearby, the mountains unfolded like bodies reclining. Farther away on the horizon was a hazy purple silhouette of rocky mountains. Coming from the lush green Appalachian hills of West Virginia, I was stunned by this landscape. It was endless rocks.

Soon we were reminded once again of the exclusivity that defined this culture. A green highway sign like the one that greeted us on the road to Mecca appeared over a lane to an exit.

FOR NON-MUSLIMS

Travel into Medina was also restricted for non-Muslims.

Safiyyah and Samir were asleep as we entered the outskirts of Medina, but Shibli was awake. I propped him on my lap to see the distant lights of the city. At this moment, we pulled over quickly into a gas station, a silhouette of the mountains in the horizon. LAND AND AIR PILGRIMS RECEPTION CENTER, the sign read.

"We're pilgrims, *jaan!*" I whispered to Shibli, invoking the Urdu word for "life," an honorific similar to "dear."

I read a sign with a missing *s* in the middle of it:

THE GOVERNMENT CUSTODIAN OF THE HOLY MO QUES
HAS THE HONOR TO SERVE YOU.

I remembered a journalist friend joking about the Saudi government's choice of the word *custodian* to describe its role toward the mosques of Mecca and Medina. Coming from a culture where the word has become the politically correct alternative to *janitor*, it struck me as a little odd too.

Although we hadn't entered the city, it felt as if we'd completed our hijrah. "Al-hamdulillah, we've reached Medina." I could feel the pulse of the prophet's mission to create a better society as I followed in his path. But I realized that I hardly felt the resonance of his life in our modern-day Muslim world. It seemed as if we had lost the passion for a better world that drove the prophet to risk death to realize his vision. It seemed to me that there were great lessons in the birth of Islam that our communities would do well to learn.

The moon hung over our left shoulders as we pulled out of the reception center and into the city of the prophet's first community. I was curious to find out what I would feel there.

THE CITY OF ILLUMINATION

MEDINA—We entered this city, like the prophet, in the dark. Before Muhammad arrived there, the city was called Yathrib. The prophet entered the city after sunset, and after his migration it became al-Madinah, "the City," because it became the model for the ideal Muslim city. It also became known as al-Madinah Munawwarah, "the City of Illumination." Most of all, it was known as al-Madinah al-Nabi, "the City of the Prophet."

Far away from the persecution he had faced in Mecca, the prophet was able to build the first house of prayer, in a small town outside of Medina called Quba. The mosque of the prophet was built in the center of Medina. That was where we were going—to Masjid al-Nabawi, the Mosque of the Prophet. It was a simple place made of mud bricks with a roof of palm leaves, balanced on wooden poles and illuminated with torches at night. A stone marked the *qiblah*, or direction of prayer, and a tree trunk had been the platform from which the prophet preached.

The mosque had been like a community center. Muslims met in the courtyard and discussed social, political, religious, and military matters. By the time he migrated to Medina, the prophet had remarried and in fact had more than one wife. Islamic scholars said his marriages were culturally accepted for political and social reasons (many of his wives were elderly widows). I had to admit that was something I was still trying to reconcile. The prophet and his wives lived in huts around the edge of the courtyard. The mosque wasn't just a place for prayer. Karen Armstrong, a renowned religious scholar, wrote in *Islam: A Short History* that "in the Qur'anic vision there is no dichotomy between the sacred and the profane, the religious and the political, sexuality and worship. The whole of life was potentially holy and had to be brought into the ambit of the divine. The aim was *tawhid* (making one), the integration of the whole of life in a unified community, which would give Muslims intimations of the Unity which is God." It was there that the prophet reformed the community so that it was a model for social justice. The prophet had a freed black Abyssinian slave named Bilal sing the call to prayer at his mosque. Bilal's emancipation and position of honor marked Islam's principles of equity.

Medina was the model for the kind of community the prophet envisioned, including women's participation and leadership in everything from the big issues of the day to the seemingly mundane. Importantly, the prophet created a community that was built on feminist ideals. The principle that women's rights are equal to men's rights defined Islam and the life of the prophet Muhammad. He was a social reformer who believed in justice, equity, and tolerance. Asma Gull Hasan, a young American Muslim lawyer, wrote in her book *American Muslims* that the prophet was Islam's first feminist. After all, he accepted as his first love and first wife a woman who was savvier, wealthier, and more successful in the world than he was. Few men, Muslim or not, would accept that kind of strength and worldliness in a wife. Based on the revelations the prophet received, the Qur'an gave women rights of inheritance and divorce centuries before women gained such rights in the West.

What truly touched me were the accounts of the Mosque of the Prophet in Medina. During the prophet's time and for some years thereafter, women prayed in the prophet's mosque with no partition between them and the men. Historians record women's presence in the mosque and participation in education and in political and literary debates, as well as in asking questions of the prophet after his sermons, transmitting religious knowledge, and providing social services. When the prophet heard that some men were positioning themselves in the mosque to be closer to an attractive woman, his solution wasn't to ban the women but to admonish the men. The mosque was not a men's club when the prophet Muhammad built his *ummah,* or "community." Nothing in the Qur'an restricted a woman's access to a mosque, and the prophet told men: "Do not prevent the female servants of Allah from visiting the mosques of Allah." The prophet himself prayed with women. Umm Hisham ("mother of Hisham") memorized the Qur'anic chapter "al-Kahf" "from no other source than the tongue of *rasulullah,*" the messenger of Allah, "who used to recite it every Friday on the pulpit." Zainab, a wife of the prophet, strung a rope between pillars in the mosque to rest upon when she tired between prayers. Another hadith says the prophet greeted women seated in the mosque.

There was a porch for conducting civic affairs, and each corner had a different purpose. In one corner Islam had an early Florence Nightingale, the Italian nurse who pioneered new respect for nurses in the modern day. Rufayda set up a medical tent on the mosque porch to hospitalize and nurse men and women injured in war; the prophet had delegated this responsibility to her, and it was a tradition that continued for centuries in mosque activities.

When some of the young girls of Medina showed interest in participating in the Battle of Hunayn, a place between Mecca and Jeddah that would require a journey of several months, the prophet took fifteen girls with his army to help in the war effort. Another time the prophet consulted a woman named Umm Salama during the signing of the Treaty of Hudaybiyya with his opponents in Mecca and heeded her advice.

At home the prophet empowered his wives to express themselves and appreciated independent-minded women of strong will, such as a woman named Ramlah, who rebelled against the patriarchal constraints in her life. She was the daughter of a man named Abu Sufyan, a leader in the prophet's tribe of the Quraysh and a passionate foe of the prophet. She defied her father's wish that she remain within the religion of the tribe and converted to Islam with her husband. To escape persecution for their decision, she and her husband started a new life in the African empire of Abyssinia, which was led by a Christian ruler, Negus. There her husband converted to Christianity and insisted that she do so also or face a divorce. Ramlah's options were limited: relinquish her right to determine her own religion by either returning to her father's home or remaining in her husband's home, or leave her husband and live alone. She chose to live alone. Ten years later the prophet sent a proposal of marriage, and Ramlah married him happily.

A strong woman in the prophet's life was his wife Aisha. Although her age when she married the prophet is of some confusion, she was most certainly a virgin child bride of perhaps the age of nine when she went to his house. As with the issue of the prophet's multiple wives, I didn't know quite what to make of her youth. It, like polygamy, was most certainly a part of the local culture, and feminist Muslim scholars say the prophet vastly improved the rights of women at the time and encouraged their self-expression. In the case of Aisha, she was a firecracker. She spent her girlhood devoted to her husband's theological mission, even though she was often immersed in the melodrama of reality. She battled her own jealousy over the beauty of the new wives the prophet married, such as a Jewish woman with the same name as my niece, Safiyyah. However, Aisha secured her place in history as the prophet's favorite wife when he died with his head on her lap. After the prophet's death, she related extensive anecdotes about his life to scribes in the mosque. In my estimation, she was Islam's first journalist; though her work didn't appear in anything like a newspaper, it filled the historical record. She relayed direct quotes, chronicled detailed narratives, and delivered rich political, social, and religious commentary. Today nearly half of the Islamic jurisprudence of the Hanafi school of thought, which is followed by 70 percent of Muslims, is based on the theology and jurisprudence communicated by Aisha to her students.

She became the transmitter of the fourth-largest number of hadith, or sayings of the prophet. To do so, she met with male scribes in the mosque. She also earned respect as a profound critical thinker and great expert in law, history, medicine, mathematics, and astronomy. She corrected many hadith, and her corrections became the subject of an eighth-century book on jurisprudence that is considered mandatory reading for any student of hadith.

After the prophet's death, Abdullah bin Umar, a leading companion of the prophet and a son of Umar bin al-Khattab, the second caliph of Islam, reprimanded his son for trying to prevent women from going to the mosque.

A profound Iranian Muslim thinker, Ali Shariati, appreciated the feminist spirit of the prophet and told his students:

> The Prophet of Islam, who was such an elevated personality and one before whom history is humbled, when he entered his home was kind, lenient and gentle. When his wives quarreled with him, he left his home and made a place for himself in the storage area without showing any harsh reaction against them. This behavior of the Prophet of Islam must be considered as an Islamic example, in contrast to the behavior of a supposedly religious, but in reality an abusive, man. Such un-Islamic, abusive behavior was based on an ethnic, cultural tradition. Therefore, distinctions should be drawn between ethnic, cultural customs and Islamic religious instructions. The Prophet's behavior was so humane that it amazes us.

I was reflecting on these lives in part because in Medina I was surrounded by reminders of the living history of Islam. The experience was having a profound impact on me, even though I had no idea where it would lead me. Muhammad turned to women as his political, spiritual, and religious advisers. He treated women as they should be treated. Like Jesus, he honored women. The legacy of male rule after the prophet's time was becoming clear to me: the spirit of Muhammad had been betrayed by centuries of men who instituted rules to protect their power. "The women of the first *ummah* in Medina took full part in its public life, and some, according to Arab custom, fought alongside the men in battle," wrote Karen Armstrong in *Islam: A Short History*. "They did not seem to have experienced Islam as an oppressive religion, though later, as happened in Christianity, men would hijack the faith and bring it in line with the prevailing patriarchy."

Being on the hajj, I thought about the role of women in the history of Islam. It made me sad to realize that people—most often men—had made up rules that not only defied logic but also were not theologically and religiously grounded. As women, we have to stay alert and refuse to be subject to the edicts that deny us our rights. As I was trying to reconcile my identity as a Muslim, I couldn't help but think about the life of the prophet. In what I read about his life, I saw a respect for women's status that couldn't help but inspire me. And what I saw in Medina was much more in line with what I felt about being a Muslim woman.

THE VISION OF MY FATHER

MEDINA—My father felt a special connection to the prophet of Islam as he walked the streets where the prophet once walked. Like the prophet who was moved to start a new community of believers, my father was always committed to building inclusive Muslim communities.

Since his earliest days, my father firmly believed that Islam is a religion of peace, love, justice, equality, respect, and accountability. He had long felt that Islam's principles of equity, justice, and respect apply to everybody—Muslim and non-Muslim, black and white, male and female, adult and child. Born in India during the British colonial rule of India, he was moved by expressions of human struggle. The start of World War II still echoed in his ear: "World war has started. Japan has bombed Pearl Harbor," the radio announcer said on the BBC. He still remembered the cuts of the tree he climbed to listen to Indian leader Mahatma Gandhi protest British rule. The scenes of death and survival during the great Bengal Famine of 1943, claiming an estimated three million to five million lives, were seared in his memory. He imagined himself as one of the children lying dead on the streets of Calcutta. Allah kept him alive. He asked, "Why?" In his reflections as a child, he found the answer: to serve humanity and care for his family and the community. "At home," he told me, "my mother was the leader of our family. She was the pilot of the ship we called home. She was the radar. She was the force. She was the *shakti*" (the Indian concept of energy). He said, "I learned from an early age to always respect women, their voice, and their authority."

Growing up, he witnessed highly educated men verbally and physically abusing their wives. He was struck by the double standard by which they preached ethics in the public sphere that they didn't practice at home. "I realized that the global society is male-dominated and social, cultural, and legal rules are mostly made by men to favor men," he said. To him, the prophet was a pioneer 1,400 years ago in encouraging and supporting women's rights, freedom, and social status. He was horrified at what had happened to women's rights in Muslim societies. When he heard about another "honor killing"—a father murdering his daughter for having sex before marriage or even for being raped—he wondered, "How many fathers have shot their sons for dishonor?" He was my fiercest defender for having accepted my responsibilities as a single mother, while my son's father walked away. "Many men do all kinds of nonsense, but they remain clean as men in society look the other way. Women are exploited or oppressed in both the West and the East, while we as men preach justice and equality," he told me.

When my father arrived in the United States in August 1962 as part of a faculty exchange program with Kansas State University in Manhattan, Kansas, he became involved with Muslim students who had met earlier in the year and formed a national organization they called the Muslim Students' Association. Then he went back to India, where he saw his first-born son for the first time, my brother Mustafa, and he and my mother conceived me. After he returned to the United States—this time to New Brunswick, New Jersey—my mother joined him, but immigration laws and my father's meager student wages kept my brother and me in India with our paternal grandparents. My parents moved to neighboring Piscataway, New Jersey, where my brother and I arrived in 1969.

As in most other Muslim communities in the United States, community organizing started with assembling a place for men to gather for the Friday *zuhr* prayer, the early afternoon prayer accorded the special status given to Sunday services in Christianity. In New Jersey my father helped coordinate the Friday prayer and build a Sunday school and local Islamic community organization. It became the precursor for the sprawling Islamic Center of Central New Jersey in East Brunswick, New Jersey, and many other Islamic centers in the state. The center's history gives credit to four Muslim Rutgers students and four community families for starting the Islamic Society and Friends of Rutgers University. My father was one of those students. There is no mention of my mother or any other women.

After we moved to Morgantown in 1975, my father started Friday afternoon prayers with some other men in the basement of the Drummond Chapel, a Methodist church two roads, a creek, and three weeping willows away from our ground-floor apartment in the West Virginia University faculty apartments. My mother and I could see the church from our front windows, but we were never invited inside for our Muslim prayer. My father took my brother as a teenager to congregational prayers with other men and boys in the basement there. The next year the men started gathering for Friday prayer at the Mountainlair, the university student union, where they got space for free. Again, my mother and I were never invited.

When local Muslim men rented space for a mosque in the summer of 1981, my father was thrilled. An immigrant Muslim was running a medical exam preparation class out of the room. The mosque was a room about the size of a narrow master bedroom. It accommodated four rows of about eight men. There was room for one row of women, but since it was so tight, the men said there wasn't enough room for women. It became a men's club. The Walnut Street station of Morgantown's Personal Rapid Transit System—a high-tech people mover—sat between my mother's boutique and the mosque. My mother was a block away from the mosque, but she never crossed its threshold. The first time my father saw women enter mosques was in 1988 when he went on sabbatical to Pakistan and saw women praying at the Shah Faisal Mosque in Islamabad and at a local mosque in Federal B Area in Karachi.

Over the years my father merged his work with his life as a Muslim as he continued his work with the Muslim community in Morgantown. The Hassan II Foundation for Scientific and Medical Research on Ramadan gave my father first prize for best research for his paper "Dietary Fat, Blood Cholesterol, and Uric Acid Levels During Ramadan Fasting." He had beat out scientists from the United States, the United Kingdom, Morocco, France, Saudi Arabia, Egypt, and other countries. It was one of the high points in my father's life, and my mother jetted with him to Morocco to claim the prize for their first trip together since their first years of marriage.

Over the years it upset my mother, however, that men could leave the house for Friday night study sessions at the mosque no matter how many responsibilities were left untended at home. My father would say: "She doesn't let me go to the masjid." But if my father had just thought about it, my mother said, he would have understood that her resentment toward the mosque was directly correlated with how unwelcome she felt there. "He could just walk away very conveniently, no matter how many things there were to do at home." She paused and recited a saying from India: "First, light the candle at home, and then light the candle at the mosque." In case I didn't get the point, she explained: "First you take care of house responsibilities, then make the bigger world a better place. To me, Islam is about taking care of your family first." She was on a roll, reciting another poem by a renowned Muslim poet, Allama Muhammad Iqbal.

The pious ones,
Overnight they made the masjid,
But we are sinners since centuries.

Yet my father, like other Muslim men, believed that women have a role in the public sphere. We met one of them one night in Medina when a man and his wife sat down at our table. There was a separate section for women who wanted to sit on their own, but otherwise men and women could sit wherever they wanted. I felt comfortable that way. We were surprised to find out that they were from West Virginia. The husband's name was Majed Khader, and he was the chief librarian at Marshall University in Huntington, a few hours south of us in southern West Virginia near the state capital of Charleston. He was clean-shaven and friendly. His wife's name was Husna Khader, and she was equally engaging. They had opened a restaurant in Huntington, the International Café, where Husna cooked gyros, falafel, and other Middle Eastern dishes.

Did any members of the local Muslim community harass him about his wife working outside the home at the restaurant? Sometimes the puritans try to put pressure on moderate men to keep their wives at home. "I don't care what they say," Majed said. I was surprised at his bluntness, but he had thought through carefully the rights that women have in Islam. He was on the Marshall University schedule in the fall of 2003 to teach a course, "Women in Islam." He had concluded that culture, man, and politics have altered divine law. "One of the biggest losers is women," he said. "They became the victims of gender discrimination." I was stunned that he so clearly expressed what I had suspected for so long. He was looking forward to reforms that would remove certain traditions and allow women to reclaim their Islamic rights. "Hopefully, the collective efforts of many will grant Muslim women their rights as vital members of society and help reinstate their rights revealed in the divine law," he said. He paused, as I absorbed his words, little realizing that I was on the brink of becoming part of that collective effort. He concluded: "Divine law does not discriminate."

My father nodded his head enthusiastically in agreement.

MY NEPHEW AND THE PROMISE OF
THE NEW MUSLIM MAN

MEDINA—In the great mosque expansion projects that lined the pockets of the bin Laden empire, the Saudis had built a massive new mosque around the prophet's original mosque. Excitement filled our hearts as we wove through the cavernous space of the women's section of the mosque to get to the original mosque. The new mosque was a world apart in so many ways: it was opulent, and, importantly, unlike with the original mosque, the Saudis had separated this grand mosque into a men's section and a women's section.

Samir smiled, remembering an earlier visit the day before. Two boys had wrestled in this sacred space during the Friday prayer. I had prayed in the women's section with my mother, Safiyyah, and Shibli. It was so massive I didn't even see the walls of the men's section in front of me.

This mosque stretches over a vast space with room for tens of thousands of worshipers. It's surrounded by marble and granite. The expansion includes twenty-seven new tombs and six new minarets, bringing the total number of minarets to ten. To me, such opulence seems unnecessary.

The imam leading the prayer and giving the Friday *khutbah,* or "sermon," was a disembodied voice over the loudspeaker. His voice bounced off the hundreds of glittering chandeliers that filled the mosque.

We had to enter the men's section in order to see the prophet's mosque. This was where the Saudis said Muhammad was buried. We were on our guard. Wahhabi Islam has changed even the way in which pilgrims pay their respects to the prophet. Some Muslims don't perform the hajj without first calling upon the prophet at his grave. But when my father performed umrah some years before, he kissed the grill around the grave of the prophet Muhammad only to see a stick in front of him, in warning. Religious officials who police the country were standing beside the grave to check the behavior of pilgrims. Kissing the grill around the prophet's grave was a violation of Wahhabi tradition, which rules that such behavior represents the worship of a human being, something disallowed in Islam. For my father, his gesture was simply a sign of respect.

My mother had decided to stay at the hotel. A relative had told us the latest hajj gossip: word was that you go straight to heaven if you pray in the mosque by the prophet's tomb, and there had been a stampede to get close to it.

"Four women died like flies," the relative had told us.

My mother looked at her, horrified. She wasn't about to join the fanaticism. She'd heard it all before. One prayer in the holy mosque in Medina equaled one thousand prayers at home. And on and on. She didn't believe God handed out lottery tickets to heaven.

Just as we were about to step into the men's section, a phalanx of women shrouded in black nikab suddenly stepped in front of us, blocking our way.

"Stop! Go back!" they shouted at us over and over again in an Arabic we understood through the nonverbal communication of their stiff spines and firm footing. We had just missed the narrow window of opportunity given to women to visit the tomb of the prophet. We could see the entryway just beyond the phalanx of women in black, but we couldn't venture toward it. We stood hopeful for a few moments that they might let stragglers through. But the women remained firm.

Shutting out women from the place of the prophet's burial seemed to be a betrayal of the presence of the feminine up until the last moments of the prophet's life. Safiyyah, at eleven, was close to Aisha's age after she had married the prophet. But if Aisha had lived today, her full and vibrant scholarly life would have been constrained by barriers that she never experienced at the time of the prophet.

As we stared at the phalanx of women in black, I realized the disappointing truth: we were banned from visiting the place where Aisha and great women in Islamic history once ran their professional lives. Dejected, we turned back through the women's section, past women in prayer and in conversation. Outside, we found my father in his white cotton pajama and kurta. He bent slightly with a crooked back to push Shibli in the Sears yellow plaid stroller I'd brought from Morgantown, its frame built slightly low for even our short statures. The marble exterior of the mosque was mostly empty, a steady stream of pilgrims and locals drifting in and out of the mosque but not overwhelming it. Surely, there was space inside for us. But, no, only men were now allowed to enter the prophet's original mosque.

"Samir, you can go with me," my father offered.

"What about Safiyyah?" Samir asked, looking up at his grandfather.

"She cannot go. Women cannot go with the men."

"But she is a *girl*," Samir insisted.

"No, she cannot go."

"That's not fair. If my sister can't go, then I'm not going. I'm only going if my sister can go." It was inconceivable for Samir to claim a right denied his sister just because she was a girl. Samir and Safiyyah had a relationship built on equality. If he got a Tootsie Roll, she got a Tootsie Roll.

"Why should I be allowed if my sister's not?" he said.

If he went and Safiyyah was denied entry, he knew she would be sitting in the hotel room unhappy. He didn't want that. Standing outside, barred from entry into the prophet's original mosque, I had an epiphany. So often, Muslim men—just like men in so many other cultures, societies, and religions—accept the entitlements that they receive by virtue of their gender, but men have the choice to reject those privileges unless they are also granted to women. In America white men got the right to vote years before women got the right to vote. And then black men got the right to vote years before black women—or any women for that matter—got the right to vote. Nine-year-old Samir faced the same choice Muslim boys and men encounter every single day: they are served dinner before women, given jobs denied to women, and, yes, allowed to enter passageways banned to women. Samir refused to walk through the door and receive the divine blessings of praying at the prophet's tomb if his sister was denied that same right. The Muslim world needed more men like him.

The sun was pouring its late afternoon rays upon us. We were hot, not to mention exhausted from the travel. Safiyyah was dejected, just as so many women have been in the face of inequity and discrimination. She crawled into Shibli's stroller with him as we walked back to the Medina Sheraton, my father's and Samir's kurta billowing in the wind. "Maybe girls are allowed during the *men's* hour," I offered, always trying to figure out a way to bypass rules that just don't make sense. "They're not," my father said, definitively. Safiyyah pondered this possibility silently as we dragged our feet home.

"Ask!" Safiyyah said back at the Sheraton.

To his credit, my father left to ask one of our leaders.

When he returned, Safiyyah sat on the edge of her bed, giving him no pause. "Did you ask?"

"They said, 'Girls are not allowed during the men's hours.'"

"That's because they marry them that young!" I said in anger.

This was the first moment that Safiyyah had stared in the face of strict segregation. She didn't like it. And I too resented it. We prayed freely at the Ka'bah, but there were limited hours for our entry in the mosque of the prophet, where he himself had allowed women to freely enter. The modern-day Saudi guardians of the mosque wouldn't even give eleven-year-old girls that privilege, not even for a few moments out of the day.

I realized that the fight for women's rights in Islam, just as in any other religion or society, isn't a gender battle of men versus women. Safiyyah and Samir helped show me that it was much more complicated. It was a battle between those seeking equity and those preserving the status quo.

Safiyyah had a friend beside her—the kind of Muslim brother who would bring about Islam's renaissance in the modern world. This experience gave me the resolve to fight this oppression, not in Saudi Arabia, where repression could have me killed, but in the West, where I wanted to help create a new reality for young girls like Safiyyah who were learning to claim their full rights in the world at large but were denied their full rights in their Muslim world. I didn't know how, but I knew that I had to do something to try to change our communities so that Safiyyah didn't grow up with her Muslim community alienating her from Islam, as mine had done to me.

That night in room 214 of the Medina Sheraton, Safiyyah wrote with appreciation about her brother's position when it became obvious that he, but not she, would be allowed to see the prophet's grave. "My brother could have gone in, but he was kind and said he would not go without me."

FAITH

MEDINA—Our room in the Medina Sheraton became a place of reflection.

"Do you believe in Allah?" I asked my family. "I don't know if I believe." I wasn't posing the question as a flirtation with atheism. I most certainly believed in a higher power. But I did doubt the presence of a being to whom I could ascribe human qualities. This pilgrimage made me wonder about my faith. Throughout my girlhood I had believed unequivocally in the concept of God. Until I hit high school, I believed firmly in a God in the way *He* is described with human attributes. But then as I battled shattered expectations in love I wondered how there could be a God who would so disregard a daughter's prayers for happiness. Michael Wolfe had told me that the hajj made him wonder about his faith. "That's the point," he had said. "You're just coming closer to the purpose of the hajj."

After thinking a while, Samir answered my question. "I believe in *Allah!*"

"Why?"

"There's so much evidence of Him. There are so many pages in the Qur'an. There are so many sermons from the prophet Muhammad. It's hard for me to think somebody made all that up. He was testing people to be good."

"Oh."

"Even though you don't know if God is for real, he told you that you should go around the Ka'bah seven times. People wouldn't do that if they didn't love him and believe in him."

"Maybe they're just stupid?"

"Two million people couldn't be stupid."

"There are hundreds of millions who aren't there."

"Maybe they can't pay."

"Maybe."

"And what about nature?"

He had me there. I found my greatest faith in nature. When I saw Shibli for the first time, I believed in God. When I saw a bird soar through the sky, I believed in God. When I saw the moon, I believed in God. It was the rest of the time that I doubted the existence of God.

"This is not a new doubt," my father said, the scientist in him coming out.

Not only were we confronting spiritual questions, but we were beginning to have doubts about our physical ability to complete our pilgrimage. Outside our hotel every day we saw men with "Bin Laden Group" on their work uniforms. As part of the contracts it had won in Mecca and Medina, Osama bin Laden's family had built one of the world's largest car parks below the mosque in Medina, and its workers filtered in and out of it beside our hotel. It added to the sense of menace.

In room 214 of the Medina Sheraton, my family was fallen from exhaustion and sickness. It wasn't just the physical challenge of late nights and constant travel, but the hajj was taking its toll, I thought, on our spiritual beings. After all, our lives, like most, were caught up in the momentum of worldly pursuits, from our jobs to the children's schooling. We paused to pray, meditate, and reflect, but we had rarely immersed ourselves so continuously in spiritual journey. Crumpled pink tissues littered the floor of the hotel room, a health hazard in itself. We were coughing germs onto each other in that room we shared.

I emerged one night for a lecture given by our group leader, Sheikh Alshareef, about the next stages of the hajj. He held court on a floor of the Medina Sheraton with an air of both solemnity and humor. We were getting ready for the chaos that unfolded in a tent city called Mina outside Mecca. We were headed there at sunrise. Although we had been in Saudi Arabia for days, the hajj truly began the next day.

Sheikh Alshareef knew about the violence that can accompany the hajj because of a ritual in Mina called *jamarat*, in which pilgrims throw stones at symbols of the devil. During the hajj in 1996, he was waiting at the jamarat to throw his pebbles. About thirty minutes before *zuhr*, the early afternoon prayer, a wave of people tumbled upon him, he said, and they all fell down like dominoes. A friend asked, "What shall we do?" Sheikh Alshareef studied the crowd, saw that they had no options for escape, and advised his friend that they should proceed with faith in the process of the rituals. "Let us go and throw our jamarat." As sirens blared and helicopters rocked the air above, he lost his friend and found himself leaning back to back against a woman he'd never met before, something he would normally never do. He thought about a saying of the prophet Muhammad that people would be naked on the Day of Judgment. Aisha had worried about lustful thoughts passing between the men and women and asked her husband, the prophet, "Won't the men and women look at each other?" The prophet had replied, "Aisha, the issue is more severe than that." The dangers were so intense that the usual issues of sexuality dissipated. Instead, something stronger gripped their souls: terror. "I see the fear in people's eyes at jamarat," Sheikh Alshareef told us, looking intently into our eyes.

Another real danger was flying rocks. For this, Sheikh Alshareef recommended gas masks. "Gas masks make good eye protectors," he told us. I scribbled the insider's tip into my notebook.

I carried his warnings back to room 214. My mother's eyes widened just so slightly in horror. The tales of the boiling sun of a city called Mina, the hike to a place called Arafat, and the stampedes at the jamarat convinced my mother she should bail out with the children. To her, no ritual is worth risking life. A sister-cousin later told us about how the flames of one of the fires that had rampaged through Mina came close to her family's tent. Her father had kept insisting, "I will die here. It doesn't matter." But our sister-cousin dragged him out. For my mother, the purpose of life is to serve humanity, not discard life in the pursuit of rituals. Instead of endangering the lives of her grandchildren, she had an idea: Mina was on the outskirts of Mecca. She would stay in the Mecca Sheraton with Safiyyah, Samir, and Shibli while my father and I camped out in Mina.

"Mom!" I yelled. "You can't quit now!"

I realized later that I press on the accelerator in my own life because my mother seems to have her foot on the brake. At the time, I was irritated at my mother for being so cautious, but I realized that there is a delicate balance between caution and risk-taking, between holding on to the security of the familiar and having the courage to grow.

To appease my mother, I called the Mecca Sheraton, hoping, of course, that it was booked up. No such luck. It had rooms. But my mother didn't insist. Instead, she pulled one of our new Medina-bought prayer rugs out of a bag and unfurled it to offer prayers that the children would emerge unscathed. I pulled out a white terrycloth bib that had somehow landed in our clean laundry in Mecca. It had an image of a smiling baby Mickey Mouse flying a kite with a baby Pluto, his tongue wagging eagerly and his tail upright joyfully. I flipped it inside out over Shibli, the back all-white side on his chest. It was the closest I could find to an ihram, the sacred toga my father and nephew were wearing. Shibli's involvement was so important to me on this hajj. He represented the faith in divine forces in me. As we got off the elevator downstairs to board our bus, Sheikh Alshareef's wife spotted Shibli, and I could feel a smile creep across her face even though I couldn't see it behind her veil.

"Baby ihram," she said, appreciatively. We piled into our tour bus, without a clue what lay ahead, and began our official hajj.

PART THREE

MAKING THE PILGRIMAGE

February 2003

SMALL ACTS AND LARGE LESSONS
IN THE WOMEN'S TENT

MINA, SAUDI ARABIA—It was an arduous journey through the night to our next stop. A Saudi government official counted and recounted the pilgrims on our bus at a stop labeled the Ministry of Pilgrimage Pilgrims Departure Control Center. The dawn arrived while we sat on the bus. The blaze of the morning sun lit Shibli's face with a beautiful radiance. As we pulled into our next transit point, I couldn't help but feel as if I were rolling into a tent city to God. We had visited this place days earlier when we were in Mecca, and it was just acres of barren valley. It sat about three miles outside of Mecca. As we drove into Mina, I saw that the Saudis had transformed the place with miles upon miles of pitched white tents. That was where we were going to officially start the hajj.

There we were supposed to praise God and evaluate our lives. The trip had already had a profound impact on me. After all our discussion, I was surprised that we had gotten out of room 214 of the Medina Sheraton, and I was curious about what the actual hajj would reveal to me.

As we got off the bus I saw an even broader representation of the Muslim world. Group A and group B had converged here from our tour group, with new faces and new stories. For instance, as we got off the bus a young man with a short beard and glasses shepherded our luggage to us. His American accent and officious mannerisms caught me off guard. His name was Suhaib al-Barzinji. He and a partner had started a company in 1995, Astrolabe, named after the navigational tool perfected during the golden era of Islamic civilization. Started as a production house, it became a leading distributor of Islamic media, from Muslim rap to Qur'anic recitations.

All around me there were new faces, and I had my guard up. I didn't fully trust my fellow pilgrims, as Muslims, because I had to admit that a part of me was afraid of them. I was afraid they would scold me, judge me, and make me feel unworthy. I worked so hard in the rest of my life to succeed and do good, but somehow in my own community I never felt I was good enough. I always had a deep sense of inadequacy.

For the first time in Saudi Arabia my family was going to be segregated, but it didn't feel puritanical, just practical. Our colony had two tents, one for men and the other for women. Outside the tents, men and women mixed easily and comfortably. Approaching the tent farthest away from the street, I pulled back the flaps and entered the sacred space that was the women's tent, Shibli tucked into my arms.

When I had been thinking about doing the hajj, Michael Wolfe said that my voice from the women's tent would be vital because so many of the tales from the hajj had come from male pilgrims who couldn't cross into that sacred space.

The tent was a wide expanse of space divided into three sections, with two rows of about twelve bedrolls in each section. There were fifty-four women under this canopy, all reflecting different expressions of Islam. Over the last days I had developed a quiet bond with several of the women. Unable to stop Shibli from crying at one point during the bus ride from Medina to Mina, I had passed him over the seats to an Afghan woman with stitches over her eyebrow. (She'd accidentally run into the glass door at the Medina Sheraton.) She had rocked Shibli so fiercely I got worried, but it had worked: he quickly fell asleep. Also on the ride, Sheikh Alshareef's wife, Amber, gave Safiyyah and Samir each a box of Nestle Turtles Original. At a rest stop, a fellow pilgrim pulled a package of Pringles down for Safiyyah. Sick from her cold, it was a buoyant moment for her.

The fronts of each section in our tent had a flap tightly closed to avert the eyes of peeping Ahmeds. Because of the fires that had broken out during past pilgrimages, these tents were supposed to be fireproof. I didn't want independent verification. By the time we arrived most of the bedrolls were taken, and we couldn't find three in a row; we didn't want Safiyyah to have to sleep next to a stranger. A bright-eyed woman and her daughter had two beds together. They were an endearing mother-daughter team, often sitting together on the bus and talking quietly to each other. Seeing us trying to figure out how we could sleep three in a row, they separated and gave us their beds so that we could sleep together.

The woman was on the hajj with her husband, a dentist, and son, both in the other tent. They lived only a few hours' drive from us in the college town of Mechanicsburg, Pennsylvania. Understated and polite, they were pleasant to be around. The mother was the kind of person I appreciated. When we talked, I felt as if she was actually listening to me. So often in my community I felt as if I wasn't having a conversation but rather that I was either listening to a monologue or being expected to quickly deliver my own monologue. After 9/11, she told us, she made an effort to talk to her local community about Islam; recently she had talked to the local media to explain that the 9/11 terrorists were acting contrary to Islamic teachings.

An undergraduate at Penn State University, the daughter had emerged as a leader in the Muslim Students' Association. The college newspaper, the *Digital Collegian*, had quoted her at an interfaith evening organized with Campus Crusade for Christ in the days after 9/11 when I was in Pakistan. She defended Islam's record on women. "Men and women are completely equal in the eyes of God," she told the one hundred students gathered on campus that night. "Women have had the right to vote, to initiate a divorce, to work, and to own property for almost 1,500 years," she said. For her, a Muslim woman's covering up of her skin "is our way of being modest." She told the students: "Women should be appreciated for their minds and for who they are, not how they look." I couldn't argue with that, and I appreciated this young college student's poise as she went through the rites of the pilgrimage. The year before she had been on security detail for the Muslim Students' Association annual convention. Not long after, Reuters published a photo of her around the world protesting U.S. plans to attack Iraq, alongside tens of thousands of others in Washington, D.C. She looked thoughtful and resolute, standing firm while gazing skyward with her sign in her gloved hands and hair covered in hijab. As 750 naked women in Australia spelled "No War" with their bodies at the time of our pilgrimage, her image remained understated but biting. "US: 9/11," the sign read. "IRAQ: 24/7." Just before leaving for the hajj, she had done another panel to explain Islam to the Penn State community, and she spoke again about women in Islam, arguing that "Muslim women are not oppressed." She pointed to the Qur'an's presentation of men and women as equals at a time in Arab culture when women were considered property. She noted that men are required to dress modestly just as women are. "To Muslims, women are precious gems, and those precious gems need to be protected."

I was happy to see my mother sitting beside the Palestinian-Jordanian woman from West Virginia with whom we'd had such rich conversations in the hotel dining room in Medina. We had really connected with her and her husband. They had been so open and engaging. Her husband was in the men's tent. If he wanted to talk to his wife, he dispatched a young boy, like Samir, into the women's tent to tell her to step outside. Samir could still navigate freely in both worlds. Safiyyah couldn't.

We'd brought too many bags into the tent. Our tentmate was kind enough to let me tuck a big suitcase between her bedding and the tent wall. Her gesture was small but it touched me.

"Thank you so much," I said, grateful for every act of kindness.

We unrolled our bedding and sat against our pillows to absorb the scene. Safiyyah was buttressed on the right by my mother and on the left by Shibli and me. To my relief, our friends from the bus had settled in around us. This was a strange country, and the company of familiar faces was a welcome thing. Sheikh Alshareef's wife lifted off her veil across from us. I hadn't seen her out of nikab, and I was curious. I thought veils hypersexualized women, rather than the opposite, by making a woman alluring and those around her curious about seeing her face.

All the while, I kept seeing acts of generosity all around me. A young professional woman across from us unpacked her shampoo to share it with Safiyyah. She touched me by the way she played so lovingly with Shibli. More than anything, I saw people trying hard to get along—a virtue in today's world. The back flap of our section opened into a restroom area with portable toilets. Some of the older immigrant women cut into the lines created by those like myself who were raised with a very Western idea about waiting for your turn. When women waiting in line protested in English, their admonitions fell on deaf ears. The women cutting in didn't understand English. It was all quite comical—unless you were the one losing your place in line time after time, as happened to me one morning as I stood to take a shower.

"*Sabar, sabar, sabar,*" I said to myself, invoking the catchall Muslim call for patience. Also a buzzword on the hajj, the authorities used it to make us feel guilty if we got impatient. There were strict codes of conduct for the hajj: no fighting, no lying, no swearing, no false accusations, and no slandering. It seems to me that Muslim communities—and for that matter all communities—would be so much better off if they lived by these guiding principles all the time. Instead, in high school, in college, at work, even at the local mosque in Morgantown, I have often heard backbiting and slander that only hurt people. My parents had given me a strong ethical training, and at its core were these principles, which were Islamic but also universal. Underlying most of these principles was a commitment to truthful living, and it was my belief in this principle that made me choose to be a journalist. It frustrated me that so often my Muslim community didn't live up to these simple codes of conduct. Gossip seemed to define the community, and it led people into deceitful, not truthful, living just to save face.

As happens to people in all communities, we were tested. We were living in luxury, and that sometimes allowed us to squabble about the mundane—such as air conditioning and toilets. Cool air blew on command from huge units between our rows of bedrolls. Sometimes the cool air gusted out so fiercely that our teeth started chattering from the cold. At those times we feared the air conditioning might trigger my mother's asthma and would use the curved white handles of our hajj umbrellas to turn the air conditioning off. One night a few women bickered over whether the air conditioning was on enough. "Sisters, I don't know why you turn off the air conditioners," one woman yelled, noting, "The tent is stinky and smells."

My mother, Safiyyah, and I just stared at each other and remained silent. The women in the tent tried to keep the air conditioning on just long enough to satisfy the pilgrim.

In the simplicity of the acts of kindness I saw in my tent, I received a serious lesson: we are the accumulation of our small deeds. The tent told me that the outside world must be like the inside world: we must be kind, respectful, and considerate, and we must live by the golden rule that Jesus taught and Muhammad echoed.

THE DEVIL'S IN THE DETAILS

MINA—At moments I felt the retributive spirit that Islamic theology, not unlike other religions, sometimes creates. My mother emerged one morning from a shower, her hair wet, tangled, and loose. An elderly woman from Sudan snapped at her, "Your hajj is not accepted. You are showing your hair!" My mother stared at her, stunned, then walked away speechless.

At home I didn't pray five times a day. Here I was trying to join the congregation. For each prayer we folded up our bedrolls to widen the aisle between our beds. After one azan, I jumped up and stood in line with Sheikh Alshareef's wife and the young professional woman sleeping across from me. As I folded my bedroll back down, an elderly woman tapped me on the shoulder and scolded me. I didn't understand what she was saying. She kept pointing at the skirt of my *jubbah*, the loose gown popular in

Saudi Arabia. I stared down. Another woman translated: while I was praying, the elderly woman said, she could see the profile of my legs, fully covered in long johns, through my jubbah. I didn't get it. Wasn't she supposed to be focusing on the symbolic Ka'bah in front of her? Why was she looking at the profile of my covered legs underneath my jubbah? This kind of scrutiny made me want to run as far away from such puritanical devotion as I could get.

In the culture of Islam, as in many religions, there is a layer of rigidity that is judgmental and oppressive. This rigidity provides a means of authoritarian control to keep people in line. I'm not sure what it is that makes some Muslims exercise this type of dogmatism. I know it arises from their staunch belief in practicing Islam literally, but it seems to be a belief in the letter of the law versus the spirit of the law. Other people, like my mother, believe that you achieve piety by who you are in your inner core, not simply by outward actions. If you are a person of love and charity, you will do good. As I had seen, the men who plotted Danny's kidnapping prayed five times a day while doing so. History has proven, however, that spirituality is a much deeper experience than adhering to rituals. Jesus angered the temple leaders of his time because he didn't follow the religious law; for example, he performed miracles on the Sabbath. I saw around me in Mina expressions of alienating rigidity in religious practice. Such puritans form a wall around the essence of religion and keep away all except the select few who know just how to mind the law.

I heard evidence of the rigidity from dispatches from the men's tent. My father was standing outside the gate to our tents when a police officer approached one of the pilgrims from our group.

"Give me reel!" he yelled, demanding the film from the pilgrim's camera and enforcing the country's ban on photography. The problem was that the pilgrim had a digital camera. It had an electronic card that tucked into the camera and recorded the pictures digitally. He turned his camera to view mode and deleted the picture. That wasn't good enough.

"Give me!" the officer yelled, grabbing the camera and crushing it beneath the heel of his boot. The camera owner told my father the story in the men's tent.

"It's so stupid," my father told me. The Saudi government accepted the strictest interpretation of sharia to ban public photography, but somehow it justified displaying images of the royal family everywhere.

In our women's tent, we listened to lectures piped into our tent from the men's tent. They were often a crackle of sounds and difficult to make out. I didn't like the separation of men and women when it deprived us of equal access to information. A Saudi sheikh visited the men's tent one night. I listened intently, trying to pick up pearls of wisdom. Instead, he horrified me.

"When you greet a Muslim," the crackle of his voice said, "say 'As-salaam alaykum.'" I knew that. It meant, "Peace be upon you," and it is often used as evidence of what a peace-loving religion Islam is meant to be. "When you greet a *kafir* (a non-Muslim)," he continued, "do not say, 'As-salaam alaykum.'"

"What?" I exclaimed to my mother. "That's ridiculous. We don't wish peace upon non-Muslims?"

Kafir had become a dirty word for nonwhites in South Africa's apartheid culture, and I resented such judgmental distinctions. Sure enough, my little pocket-sized prayer book had a special greeting for non-Muslims. Prayer number 123: "Returning a greeting to a *kafir*." It stipulated that the greeting should be: "And upon you." The greeting returned to a Muslim: "And peace be upon you." The distinction disgusted me. My mother read my mind and murmured to me: "These are the things that turn you off." I didn't hear protests from any of the other women.

In the men's tent my father was challenging this teaching. Our imam, Sheikh Muhammad Nur Abdullah, the president of the Islamic Society of North America and leader of a mosque in St. Louis, didn't agree with the ruling. "Greet everyone," he told my father, "with 'As-salaam alaykum.'" That relieved me, and it reminded me again that Islam simply isn't practiced one singular way—no matter how much any one Islamic ideology insists that its path has a monopoly on virtue. From our *adab* (culture) to the way we pray, Muslims have many paths to Mecca.

SEX, LIES, AND TRUTH

MINA—Inside our tent, something remarkable happened. Shibli seemed to take his first step. It was a gentle stutter step with our hands to catch him, but my maternal eyes saw it as his first little step. A first little step was the least we could all hope to do in our own hearts, although the hajj seemed to be a major step for me because it was helping me define and clarify who I am and who I want to be. One thing I realized was that I want to live truthfully. I had to lie all through my twenties when I felt I couldn't tell my parents the truth about my love life. I realized that living a lie is like living with a noose around your neck. Living a lie is harder to do than living truthfully.

I stood between the kitchen and the tent where they had put four portable toilets. Another woman waited in line with me. She was from Pakistan but had a different air about her than the aunties I'd gotten to know from Pakistan. For one thing, she was exuberant. She would yell across the tent to try to get Shibli's attention.

"*Shhhhiiiiiiibli!*" she'd yell. Her name was Zarqa Ekram, and she lived in Atlanta, Georgia. As we stood there she asked me, "Are you married?"

I was too stunned to answer immediately. Everyone had assumed that I was married, and I had played along with the charade. Ever since I landed in Mecca, I had been living a lie. Zakia Zikria, a young, fresh-faced Afghan American pilgrim from Alexandria, Virginia, had innocently asked me about my husband while we sat amid our tumble of luggage and shopping at the Jeddah airport. "He lives in Pakistan," I had said, inventing a wedding contract between Shibli's father and me. At a rest stop between Medina and Mecca, at a mosque, a woman had said to me, "Shibli's father must really miss him."

"Yes, I'm sure he does," I answered, figuring that, yes, any father would miss his child.

"What does your husband do?" another woman had asked me. I had answered with the truth about what Shibli's father did in Karachi, without clarifying that he wasn't my husband.

It was a tangled web, and I didn't like it. I had been flipping from time to time through the pocket prayer book we'd gotten on the trip to Mecca. "Exhort one another to truth," it said, repeating Qur'anic chapter "Asr" (103:3). I felt conflicted by deceit.

I returned to the woman asking me about Shibli's father. "You're not assuming that I'm married?" I asked her.

"What do you think? I have daughters. I know life isn't that simple."

"I'm so relieved," I told her, and I admitted the truth to her. "I'm not married. I'm raising Shibli on my own with my family's help."

I felt as if I had pulled the noose off from around my neck. From childhood, I had confronted the repression of sexuality. I had learned at an early age to not talk about this part of myself because it was so taboo. In the fifth grade my mother didn't want to sign the permission slip for the school's sex education class. She relented when I insisted, and I didn't learn anything that I thought was sacrilegious. Throughout my teen years I never had a conversation with my parents—or anyone else, for that matter—about sex. When I was twenty years old and confessed to my mother that I had had sex, she told me: "Stop!" And that was the end of the conversation. When I tried to talk to her about it again at the kitchen table around the same time, she started sobbing uncontrollably. My mother told me later, "My upbringing taught me there was a right and a wrong. According to my values, dating was wrong. My world was black and white. I didn't know how to travel in the gray area. Only later did I realize that we have to have open discussions even if they are difficult." The message was clear that I was doing something terribly wrong. By having to stay secretive about my relationships, I didn't have a community to help me make wise decisions about relationships.

When I became sexually active, I began to understand the power of sex, but I didn't see why we attach stigma to adults having sexual relations. I always believed there was a better way than the way I was taught to think about sex, which was that it is wrong, dirty, and sinful outside of marriage. Then in our meetings in Mecca, I had to confront the "sin" of having a child out of wedlock.

Our group, like most, met daily for religious lectures. We took our seats in a room at the Sheraton, men and women sitting wherever they wanted. I sat in a row of seats along the side. The speaker one night was Muhammad Nur Abdullah, the president of the Islamic Society of North America. A tall man with dark skin and a calm disposition, he cut an impressive figure. He told us our old selves died symbolically on the hajj. The pilgrimage was an allegory for our final farewell from earth. We were supposed to make our final will and testament before departing. We were also supposed to pay our debts, erase ill will, resolve conflicts, seek forgiveness from others, and realign the moral compass of our lives.

"Like a new baby," he said, sweeping his hand toward Shibli, sitting on my lap, "we are forgiven of past sins. We are blessed to be here."

Sheikh Abdullah guided us to repeat after him: "Oh God, forgive us."

I whispered the words underneath my breath, but I didn't utter them from my heart. This concept of forgiveness eluded me. My father and I had an argument about it before we left Morgantown. My father believed fully in forgiveness, but I asked, "What's the point of asking for forgiveness for decisions that can't be changed? Why live with regret?" At that time, I was resisting the concept of forgiveness because so often it seemed to be attached to regret, punishment, and repentance. If I sought forgiveness for Shibli's conception, did that mean I regretted it? I didn't and never would. Did seeking forgiveness mean I believed I should be punished? I didn't and never would. It was only much later, with the help of an American Buddhist teacher named Tara Brach, that I learned that seeking forgiveness is a way to be compassionate with yourself about actions that carry emotions such as shame, blame, or hurt. With that understanding, I did seek forgiveness. I wanted to free myself from my self-loathing over the errors in judgment that had led me to love and trust a man who left me hurt, sad, and alone.

I struggled with the question of how to find forgiveness. Did I earn it? Did I punish myself for it? Did I have to be punished to deserve it? With these questions, I listened to Sheikh Abdullah. "May Allah (*subhana wa tala*) forgive us," he said, invoking another parlance in the secret Islamic code, meaning, "the sacred and the mighty." "We come with our backs heavy with sin."

Was my back heavy from sin? Truly, my back ached from the weight of carrying Shibli on my chest over the past two days of the pilgrimage. Was creating Shibli the sin for which I had to seek forgiveness? I knew I had to resolve this question. I brought Shibli close to my heart as I slept that night and felt as if I knew the answer with the softness of his breath against me. Even though I wished I had resolution, I didn't.

I decided to resolve this issue. So often, traditional society in so many faiths defines gender relations by sin and sexuality. Women have so often been elevated—or denigrated—based on sin and sexuality. Mary, the mother of Jesus, has to be a virgin. Even Islam considers her a virgin. Mary Magdalene, a companion of Jesus, is conceived as a harlot and prostitute, although modern historians dispute that portrayal. The Bible frames Hagar as the unmarried sexual concubine of Abraham. Islam redefines her as a second wife; even if that conception is accurate, she wasn't equal to the first wife as a co-wife. After all, Abraham, based on the stories of all the faiths, left Hajar in the desert while building his home with Sarah.

I felt the pressure of the weight I carried on my back for the sin of having had sex as a single woman. According to the sheikh's logic, with the hajj I would become a reborn virgin. Did I need that in order to feel good about myself? To feel pure? To feel worthy? It made me wonder why we had to associate gender relations with sin. In Mecca I realized that I needed to examine the messages that I had received about sex.

In recent years I had started thinking about issues of sexuality from the vantage point of a cultural anthropologist or sociologist. At the *Wall Street Journal,* my friends joked that I was the newspaper's informal sex reporter even though my beat was officially travel at the time. I was trying to understand sexuality outside the prism of religion. I even wrote a memo to the managing editor of the *Journal* proposing a beat on the sex industry, arguing that it paralleled major industries in sales, labor force, and contribution to national GDPs and that we weren't giving it the journalistic attention it deserved. I never sent the memo.

This led me back to reflections about my own sexuality, which had been germinating since my earliest days. Everyone has a choice: either we figure out on our own what we believe about sex, or we accept a religious authority's edicts about it. It was during that period that I got the assignment to report on the big business of Tantric sex. It was a peculiar philosophy that had me traveling from Canada to California for weekend workshops on sacred sexuality. Partly because sex sells, these workshops focused on concepts such as "sacred spots" and "divine love." My page 1 article about Tantric sex turned into a project to write a book about Tantra. As a Muslim, I ventured into Hindu and Buddhist philosophies about sacred sexuality. I learned a fundamental concept during this assignment: sexuality is a vital part of our being, and we're best served if we deal with it as a healthy part of society, not as something to repress, sanction, or adjudicate.

I also discovered that Islam has a very rich tradition of sacred sexuality. The prophet Muhammad talked about the "sweetness" of intercourse, and he dealt realistically with issues of pleasure, desire, and even frustration. Children were married when they reached puberty. I don't think it's appropriate in today's age, but child marriages at that time allowed for a codified expression of sexuality at the age when it ripens. The prophet told men not to leave their wives for more than six months because sexual tensions would get too high. Islam rejects celibacy as a way of life. And the Qur'an speaks eloquently about the concept of sacred sexuality between husband and wife.

I was drawn into exploring sexuality and gender relations because I saw the profound effect of these issues on our lives, from power, position, and social acceptance to love, marriage, and intimacy. I knew I had a commitment to speak out about issues of sexuality because my own experiences told me that this subject needs to be discussed among all religions.

And I had to think about what I was going to teach Shibli. I wanted to raise him in an environment in which he could discuss sex with me openly and honestly, in which we both acknowledged that he might have sex before he was married. I wanted to present sex to Shibli as an act of love between two human beings carried out with respect, honesty, and responsibility.

Over the ages, healthy relations between men and women have been repressed in the name of religion. Just as Prohibition didn't work in the United States with alcohol, the puritanical repression of sexuality doesn't work. In Pakistan, I saw that sexual repression creates a hypersexuality that leads men to sexual dysfunctions, such as premature ejaculation. It also creates an atmosphere in which women are killed by their brothers, husbands, and fathers for allegedly breaching the sexual honor of the family. I rejected this ethos. I didn't believe that my worth was attached to my sexuality, and I was relieved to have gotten pregnant. My baby allowed me to be honest about myself as a sexual being. I believed that as a society we need to be honest about sexuality as a part of our communities.

JUDGMENT DAY, SIN, AND FORGIVENESS

ARAFAT, SAUDI ARABIA—We were on the eve of a ritual called the Day of Arafat, an annual event that marks the most important day of hajj. We were going to a wide, open, desolate plain called Arafat, twelve miles east of Mecca. The Day of Arafat symbolizes the judgment day when we go before God to be judged for the acts of our life. There are no secrets from God. "The hearing, the sight, the heart—all of these shall be questioned," says the Qur'an (17:36).

I had heard about this moment since I was a child. Heaven or hell— our eternal fate is decided on that day. I had heard that we will walk a tightrope the width of a strand of our hair. We will be able to find our balance if our good deeds outweigh our bad deeds. We will fall into the pits of hell's fire if they don't. I vividly remembered the day when I first heard the concept of "hellfire." We had moved to Morgantown, and we were living in the faculty apartments. My mother was teaching my brother and me about Islam. Even though she had become a skeptic, my mother taught me the most about Islam, its rituals, and its beliefs.

"Stay on the straight path," she had warned us. If we didn't, a tragedy awaited us: hellfires. From that moment forward, I despised the fear that religion puts in our hearts. I wanted to be motivated by love of Allah, not fear.

On the bus ride to Arafat, I suddenly realized that I didn't know what to believe. It seemed like an auspicious day to figure it out. In a dramatic way, I was facing the possibility of God's forgiveness and benevolence. I remembered the exact moment when I realized that Allah is forgiving. It was a few years back when I lived in Brooklyn Heights, and I was standing next to the window through which on so many nights I would gaze at the full moon ascended over the steeple of St. Ann and the Holy Trinity Episcopal Church. One of my cousins had come to visit a local college with his high school daughter. I had just gone through another heartache with yet another man who dashed my hopes for love. My cousin and I had gotten into a theological argument about Islam. "But what hope is there for any of us if Allah doesn't forgive?" I asked.

My cousin looked at me gently. "But Allah *does* forgive."

"He does?" I asked, realizing that I had somehow evaded one of Islam's greatest teachings. God forgives.

"He does," my cousin gently said.

That night I called my mother and complained, "Why didn't you tell me that God forgives?"

My mother sighed. "I did, Asra. You forgot."

On the way to Arafat and my day of reckoning on this earth, I pondered the consequences of forgiveness. I mostly doubted the anthropomorphic sense of God as someone who sits on a heavenly throne, particularly as someone who would annul the blessings of hajj over something like wearing a safety pin on one's ihram. I gazed out the window at the devoted piled into open-air school buses. They spilled out of trucks like cattle. The people with the darkest skin seemed to be the most destitute. These pilgrims wore air masks to block the car fumes. My family and I sat in air-conditioned luxury buses. We passed a pristine building marked SAUDI RED CRESCENT SOCIETY. The Saudi society was the equivalent of the Red Cross in the West. Instead of the cross that marks the Western charity, the Muslim group uses the crescent, the symbol of Islam, as its mark. We drove by canvas tents where worshipers spent their nights amid litter. To fulfill the rules of hajj, we had to arrive before noon.

Our physical journey took me back in time. There was another woman besides Hajar who found herself alone in the desert outside Mecca. She was Eve. The Qur'an, like the Jewish and Christian histories, says that God made Adam and Eve as the first man and woman. The angels protested God's plan: "Will you create beings there who will cause trouble and shed blood, while we praise Your Holy Name?" Indeed, God made Adam and Eve with a soul, free will, intelligence, reason, and *fitra*—an inner nature that seeks God and is disposed toward virtue. After being exiled from the Garden of Eden, Adam and Eve were first separated, but God answered Adam's prayers and reunited him with Eve on the top of a hill called Jabal-ur-Rahma, or the Mount of Mercy, in the valley of Arafat outside Mecca. It's said that Eve spent the whole day with Adam, the two of them standing modestly, thankful and awed in worship of God. For that reason, the hill is called Arafat, meaning "recognition" or "knowledge" in Arabic.

Unlike Genesis, the Qur'an doesn't single out Eve for blame in tempting Adam toward sin. "They both sinned," the Qur'an says. But interpretations of Islam portray women as temptresses who can doom a man with just a glance or a strand of hair askew, hence all the prescribed restrictions to protect women's honor. The story of Eve underscores the issues of sin and redemption that Muslim women face in a religion that defines every aspect of their lives, from the way they dress to how they have sex.

Like other women wracked with guilt, I'd had to grapple with the question of whether my sexuality had led me astray. My religion attaches sex to marriage, and since I was unmarried when I conceived, I had to wrestle with the issue of religion, sex, and sin. I saw that women are the ones who are condemned for sexual activity, while men are allowed to walk away. I concluded that the doctrines of religion shouldn't set the parameters for our bodies and our hearts. All adults, including women, should be able to make these intimate determinations for themselves. That way, we can be free of all the guilt and shame that too often mark the lives of people of all religions but do nothing to promote healthy societies.

MY JUDGMENT DAY

ARAFAT—On the ground in Arafat, the prophet Muhammad followed the path of Eve, riding a she-camel in the afternoon of his hajj as far as the middle part of the Arafat Valley on the Mount of Mercy.

This is the main hajj rite, the day known as Yawm al-Wuquf, the Day of Standing. The prophet Muhammad said, "Hajj is Arafat." The Qur'an (42:11) says, "The Beneficent . . . [rose over] the [Mighty] Throne, over the seventh heaven [to us], and he only comes down over the first heaven on the Day of Judgment." We were supposed to stand until sunset in the same valley where the prophet stood, speak to God, and ask for mercy for past sins, as if our judgment day had come.

There, in the year 632, the prophet recited his farewell sermon, known as the Sermon of the Farewell Hajj. He said in that sermon that no one is superior to any other, a message lost in the schisms that now divide the world's religions. In the Uranah Valley of Mount Arafat on the ninth day of the month of Hijjah in 10 A.H., the prophet seemed to know he was going to die. There, in his farewell sermon, he said: "Oh People, lend me an attentive ear, for I don't know whether, after this year, I shall ever be among you again. Therefore, listen to what I am saying to you carefully and take these words to those who could not be present here today."

He spoke of the golden rule: "Hurt no one so that no one may hurt you." The essential message was one of equality in duties, rights, and obligations. "Oh People, listen to me in earnest. Worship Allah, say your five daily prayers, fast during the month of Ramadan, and give your wealth in zakat. Perform hajj if you can afford to. You know that every Muslim is the brother of another Muslim. You are all equal. Nobody has superiority over another except by piety and good action."

This day marked an important moment in the history of women in Islam. The prophet Muhammad's sermon included a message for the empowerment of women. "Oh People, it is true that you have certain rights with regard to your women, but they also have rights over you," he said. "Do treat your women well and be kind to them, for they are your partners and committed helpers."

Our bus pulled up to a lane lined on both sides with tents. We slipped behind one entryway. The men slipped into the first tent, and the women continued to a second tent. There were mattresses everywhere, as if we were there to nap. I was confused. But as with so much else on the hajj, I tried to follow the spirit of what we were supposed to do.

I had always had trouble accepting the concept of sin. As in so many religions, it was used in Islam to instill a spirit of fear and punishment that I intuitively rejected. For example, when my mother's hair once again spilled out of her hijab, Samir asked her, with incredulity and slight annoyance at having to protect her again from the rigidity around us, "Do you *ever* cover your hair?"

I resented having to live with such fear because of this amorphous concept of sin. Before we left for New York, I told Samir, "We're supposed to wipe away our sins with hajj, but that's not a problem for you and Safiyyah. You both haven't sinned."

Samir shot me a quick look. "Nuh-uh! I've sinned zillions of times."

"When?" I asked. I couldn't imagine that he had anything to confess.

"The time I hit Safiyyah."

"When did you do that?" I asked, surprised. Samir and Safiyyah got into the usual sibling tussles, but nothing violent.

"The day Safiyyah sinned! She pushed me into the rocking chair!"

Indeed, it had been a toddler tussle that had given Samir a permanent dimple on his right cheek.

Even in her innocence, Safiyyah recognized this day as the most important of hajj. "My cousin Shibli stood with me and prayed," she wrote in her journal. "We saw a pretty butterfly." She appreciated the unity of the day. "Today was a beautiful day with everyone remembering God! together."

Outside under the sky, with the earth beneath his feet, my father stood contemplatively under a tree. When the men complained about the air conditioning not working in the tent, my father retorted: "There's a nice breeze. Sit under the tree."

It was as if my father were in heaven. Throwing his hands into the air, he spoke directly to Allah. Tears welled up in his eyes. His voice quivered. He asked for blessings for everyone. "Please forgive me, Allah, the sacred and the mighty, for anything I have done intentionally or unintentionally. Please make me a good person to serve humanity as long as I am alive."

He couldn't stop weeping.

It's said that the prophet proclaimed, "There is no day on which Allah frees more of His slaves from fire than the Day of Arafat, and He verily draws near, then boasts of them before the angels, saying, 'What do they seek?'"

I stood outside with Shibli on my chest in his baby carrier.

During the hajj of 2002, Saudi Arabia's leading cleric took the moment of the great sermon on Arafat to criticize the "pillaging Jews" in Israel. What would today's sermon be? I would never know. I tried hard to hear it. It was just a crackle over the loudspeaker system. I continued to stand with my son. Shibli faced out, his back against me.

We prayed, as the prophet did, with *wuquf*, or devotion. People around us were also praying, begging for forgiveness of sins, weeping, in the spirit of Eve and Adam. The prophet said that the prayer of this day was the best of all. The women around us were in various stages of prayer—and exhaustion.

Standing outside with me, Safiyyah asked, "Did we pray enough?"

As Shibli fell asleep, I lifted him out of the carrier and lay him beside me on one of the many mattresses that lined the courtyard. In his sleep, Shibli kicked the pen out of my hand as I was writing. Nearby, a pair of white shoes with the Wal-Mart "No Boundaries" label sat outside our tent. A woman slipped into them and stepped inside. This Wal-Mart line has made me wonder about boundaries ever since I first saw the label. Should we live with no boundaries? In Arabic, it would be *la hudud*. In my life I ran up against boundaries and crossed them. I hadn't yet resolved my thinking about boundaries—their worth and their cost.

Inside the women's tent my mother wanted her judgment day to end. It was so hot that she couldn't concentrate. An African American nurse from Boston lounged on her mattress and complained, "What about the air conditioning?" She listened as a pilgrim joked that the air conditioning would turn on just as we left.

My mother stepped outside into the courtyard to get some fresh air. She was more than angry about the heat. She was angry with God. "You gave me so much misery. Since you have given me so much misery, help me out of it," she told God.

She asked for Shibli to have a good father. She prayed for peace of mind for everyone in her family. Then she went back inside and lay down, exhausted.

Before the sun set, we were ready to depart to follow the centuries-old trail that Eve took with Adam to Muzdalifah, a place between Mina and Arafat. The movement of pilgrims from Arafat to Muzdalifah is called *ifadah*, or overflowing. Piles of discarded rice and meat stretched along the side of the road as we left. It was filthy outside, with the trash of the day all around us. Humanity seemed so flawed on a day that was supposed to be a bridge to the divine.

On the way there the prophet Muhammad kept saying, "Labayk. Allahumma labayk." Here I come. At your service, Oh Lord, here I come. The people were chanting the same. I couldn't know God's judgment of me, but I felt clean as I left my judgment day.

AFFIRMATION

MUZDALIFAH, SAUDI ARABIA—When he got to Muzdalifah, the prophet asked his companion Bilal, the emancipated Abyssinian black slave, to give the call for prayers. He said the same *maghrib* (sunset) and *isha* (night) prayers that we now said.

It was considered a blessed night there in the place where the prophet rested. The stars engulfed us in their splendor. Pilgrims were sleeping everywhere, not on cool desert sand but on asphalt. The Saudis had paved acres of the land there to make a massive parking lot in the holy land. My father and Samir wandered through the parked buses and cars and sleeping people to collect pebbles to throw the next day at pillars that symbolize the devil. They were a wonderful image to see: Samir in his white, my father in his white, the two of them separated by two generations but united in purpose: collecting stones for the three days of stoning.

Late that night I sat on the bus with Shibli in my arms and talked to one of our guides. Somehow the conversation turned to Shibli, and the guide asked me about Shibli's father. I marked the first day after my judgment day on earth by telling our guide the truth about Shibli's father. He didn't judge me but simply asked: "He wasn't a practicing Muslim?"

I didn't know how to answer that question. It wasn't my place to judge. He certainly hadn't treated me with kindness, even in those vulnerable months when I was carrying his baby. But I didn't know what beat in his heart. I hesitated, and the guide filled the silence with a prayer.

"May Allah give you a good man."

"Thank you."

His prayer made me realize that I could have a husband—that I was worthy of finding happiness in a marriage.

I curled into a seat on the bus and slept with Shibli close to me. The dawn broke, and we stirred awake.

THE TRUE FACE OF SATAN

MINA—The masses seemed to flow everywhere around us as we rode our tour bus to our tent city. An old man and his wife lumbered through an empty ravine. A flag of Pakistan floated by us. A mother in a burka shepherded her brood, a girl on her shoulder and another girl with a white scarf on her head. I saw an African woman nursing her child. They were strangers, but I felt so connected to the mothers, each leading her family through this journey. What it said to me was that I now had a new definition as a mother. I looked at the women and felt a kinship. As much as the rituals of the hajj were supposed to connect me more to my faith, I was more moved by the people around me. They were affirming a spiritual connectedness between me, my son, the world, and history.

A tidal wave of emotions struck me among this mass of humanity. We passed a pilgrim with a white umbrella with BANGLADESH on it. Another umbrella read EGYPT AIR. Another, AIR INDIA. A man waved a flag: HAJJ GROUP SRI LANKA.

A girl balanced a white plastic container on her head. We were all headed to the same place on this morning, to stone Satan. The trash around us horrified me. It didn't seem like a sign of respect to either God or the earth. I wondered how the devoted could be so careless. We were going in the direction of the qibla—the direction that points to Mecca. Dark-skinned men in yellow jumpsuits picked up the trash. They were all from either the Indian subcontinent or Africa, it seemed. The old man and his wife we saw a moment before had found a seat on the roof of a van that drove by us. They looked beautiful.

In our bus we passed through the valley where warriors on elephants had gathered to proceed to Mecca to destroy the Ka'bah. When he did the hajj, the prophet Muhammad instructed pilgrims to move quickly through there. As we continued over the bumpy road we passed by a sign that read MUZDALIFAH ENDS HERE. It marked our passage from the valley of redemption to the summit on which we were to reject sin.

A sign beckoned us to the place where we were supposed to confront the devil. MINA STARTS HERE, it read. The streets were so crowded that our bus had to stop. We walked the rest of the way to our tents in Mina, not a long walk in distance—just half a mile—but a long way in perseverance. The scene in front of us was disgusting: we walked in a packed stream of pilgrims through a path of dirty water, trash, and sleeping pilgrims littered beneath a bridge named for King Khaled Ibn Abdul Aziz, the wealthy Saudi family heir who ran the county from 1975 until his death in 1982. I realized that, to me, religion is about helping others in need before it's about praying five times a day. I was in the midst of a religious pilgrimage, but it seemed to me that the best thing we could do in the name of our religion was to stop the rituals and help the people around us.

The scene under the bridge was testimony to the worst that happens to humanity in the name of religion. It was so bad that the Saudi government even warned about it in a health brochure that an English expert clearly hadn't edited: "Dear Pilgrims, Laying under bridges and setting on the footpaths is an uncivilized behaviours, and doing so causes you many risks. so do not exposure yourself for risks."

Of course, the pilgrims wouldn't have slept under the bridge if other accommodations had been available to them. Another one of the Saudi government's tips seemed to defy reality: "Crowding is an important factor for transmission of meningitis, try to avoid crowded areas." Trying to avoid crowded areas on the hajj was like trying to stay dry in the ocean. The dangers became apparent to me in one swift moment. The crowd started to crush me, pressing me toward a wall of squat buildings on my right. We were going forward. Samir was troubled by seeing the people close in on him. There were slippers everywhere too. People had lost them and been unable to retrieve them in the crowd. Our guide, Sheikh Alshareef, had warned us: no matter what, don't stop to pick up lost shoes. Let them go. Otherwise, we risked being trampled in the press of the crowd. Stopping for shoes was an invitation to death.

Samir said quietly to himself, "Allah, please don't let me be trampled." He looked at the packets of water littered on the ground. They looked useless to him, but then he saw pilgrims dousing their heads with the water. "Oh, I get it," he said to himself. It was so hot that the government of Saudi Arabia doled out packets of water to cool pilgrims down.

Shibli squirmed in his carrier on my chest. It was increasingly hot, and I jabbed my hand into the crowd like a linebacker, trying to protect Shibli. Safiyyah recalled later: "I almost died because I was getting crushed."

Women and men yelled at me in Arabic. I couldn't understand what they were saying. "What?" I exclaimed.

"They're saying, 'Put your baby up!'" Sheikh Alshareef's wife translated. "How?" I exclaimed. Suddenly, a handsome young Egyptian American man broke stride beside us. "He is my friend," Sheikh Alshareef's wife said. "He can carry the baby!"

I hesitated. I didn't know this man. I could lose my son in this crowd. To avoid the danger of choking on an ID tag's cord, I didn't even have Shibli wearing his ID, which identified him as "Card Number 34" in our tour group. I had a badly photocopied map of Mina folded into my pocket with phone numbers beside Arabic script I couldn't read. The only instructions in English: "In case of lost," we had to look for Mena Square 49 under the King Khaled Bridge or call a Mr. Arafat on his mobile phone. Of course, I had no idea how to get to a public phone. My only other option seemed to be to find refuge in the open door of a smelly men's restroom beside me. It was one of those perilous moments that mothers have faced since the beginning of time. I chose to take the calculated risk.

I handed Shibli to the young man and tried to stay close beside him. We navigated gingerly but forcefully through the crowds. Shibli rested on the man's shoulder. I felt so grateful, yet remained worried. I couldn't risk losing my son. Finally, we took a turn out of the crush. We proceeded quickly to our tent, where the young man brought Shibli safely into my arms. "Shibli!" I said, grateful. We tumbled into the women's tent, where, to my shock, tears awaited us from our fellow pilgrim from Mechanicsburg, Pennsylvania. She had gotten separated from our group in the crush and arrived just moments earlier. She had been weeping because she was so worried about Shibli's safety in the crowd. She epitomized the best of what religion teaches—compassion for others. In pursuing ritual, some people would have stepped on my head and Shibli's too, and somehow that made no sense. But this woman, a virtual stranger, had tears running down her face because she was worried about my baby's safety. It's said that the hajj is supposed to teach us spiritual lessons in the most unusual of ways. I smiled at the woman and offered her a gesture of gratitude. I dropped Shibli into her arms.

Meanwhile, Safiyyah fell onto her bedroll to float into the safety of her dreams. All she wanted to do was sleep and turn the air conditioning on—really high.

This trip revealed to me what is good and evil. To me, evil is social injustice, discrimination, bigotry, and intolerance.

In the tents of Mina I listened to lectures piped through a sound system from the men's tent and heard the story of Satan, or Shaytan in Arabic. The Qur'an says that when God made Adam and Eve, he educated Adam with knowledge about the wonders of the earth. The angels were wary about a human having such abilities, so God tested the angels' knowledge of nature.

"Tell Me about all of this," he ordered, speaking of the earth.

The angels, who lived in a different realm, admitted defeat immediately. When Adam explained animals, plants, and the world to them, the angels conceded that humans were superior to them in intellect. God had the angels bow to Adam to pay their respects. Some Muslim scholars interpret this as God's support of human rights. The angels bowed, and bowing with them were creatures called jinn that God had also created. They were surreal creatures that had followed me into my childhood. My mother used to tell me stories about jinn. They were like the spirits of haunted houses, only I never felt afraid of them. There was one jinn that we were particularly supposed to fear: Iblis. He stood with the other jinn in Paradise, but he refused to bow to Adam. God asked him why.

"I am better than him," Iblis declared. I was later told that this statement made Satan the world's first racist. Racism was unacceptable in God's world. Angry, God turned Iblis out of Paradise, according to the Qur'an. But Iblis had a wager for God. "If you give me time, I can corrupt [your precious humans], and in the end you will find most of them ungrateful to You." God accepted the challenge and gave Iblis free reign over the world until the Day of Judgment. Iblis vowed: "I will attack them from their front and back and their right and left, and I will create in them false desires and superstitions." God set only one rule: Iblis wouldn't have any power over anyone who sought God's protection. Iblis made the deal. His name became Shaytan, meaning "to separate." Some of the jinn went with Shaytan and took the name Shayateen, or "separators." By the standards of "the straight path," I had separated. I had become a bad girl of Islam.

I had committed many of the so-called vices, and by the strictest standard my worst trait was not being ashamed of my errant ways. To me, what I had done was a part of life, and I wasn't going to punish myself for it. I also didn't live in fear of punishment from God. Fear underlies so many Muslims' belief. We are always told, "Fear Allah," but I didn't fear God. And for that matter, I didn't want to fear God. I thought the dance of love and fear of the divine was an unnecessary drama. I also didn't *hate* the devil. I didn't even believe in the concept of a devil with human attributes. We all had a dark side. But I had come to believe that we allowed ourselves to be more positively transformed if we *accepted*, not despised, our dark side. I had read somewhere that it was just like the way we accepted gentle pain during stretching exercises. These feelings didn't put me in a repentant space as we entered day four of the hajj, the day when we were supposed to face the devil.

On that day and the two that followed, we were supposed to throw our stones at the devil in a ritual called *ramy*. The devil was symbolized in three stone pillars: al-Jamara al-Kubra was the tallest pillar, al-Jamara al-Wusta was the middle pillar, and al-Jamara al-Sughra was the smallest pillar. As sacrilegious as it sounds, I couldn't help but feel like Goldilocks facing Papa Bear, Mama Bear, and Baby Bear.

The prophet said that when Abraham wanted to do his rites of hajj, Satan blocked his way. Abraham threw seven pebbles at him, and Satan sank into the ground. Abraham proceeded to the second pillar and threw another seven pebbles at Satan, and again Satan disappeared into the ground. Blocked yet again at the third pillar, Abraham again threw seven pebbles.

The devil wasn't the only thing to fear here. In 1998 a stampede in the plain of Mina killed 118 pilgrims as large crowds assembled near a bridge on their way to the devil-stoning ritual. I was going to go alone against the devil. I left Shibli at the tent with the women there, his first time spent with strangers since he was born. He was purer than any of us, I believed, just descended from heaven. Safiyyah was more practical about the virtue of leaving her cousin in the tent. "Shibli might accidentally get hit by a rock!"

Sheikh Alshareef and his wife, Amber, were leading our group and trying to play it safe shepherding us through the pilgrim rush-hour traffic, as we had done in Mecca. As we walked together through the crowds of pilgrims, I absorbed the scene around me. The modern day mixed with mythology. Not far from the place for throwing stones at the devil sat a rotisserie chicken restaurant that went into Safiyyah's journal as the "best chicken in the world" when she got a taste on another day. I kept pace with Amber, enjoying her conversation. She was fully cloaked in her nikab with its face veil, and I was wearing my hijab only to stay out of jail, but I felt a connection with her that I enjoyed. What she and her husband showed me was that we can be diametrically opposed to each other on important theological and ideological questions—the veil, polygamy, interpretations of sharia—but still get along. We can still respect each other. I also learned that we don't have to abandon simple rules of decency, civility, and friendship even when we differ profoundly on issues as important as God's so-called law.

We paused near puddles and groups of pilgrims huddled together in casual conversation as Sheikh Alshareef went off on a reconnaissance mission. He returned and said that we needed to wait a while for the crowds to thin. I was in conversation with Amber and didn't mind the delay. We watched my father talking enthusiastically with another pilgrim. I shook my head. Amber didn't need a translation. "I tell my father too, 'Can you stop talking?'" I laughed, hearing my own voice in hers.

Then it was time to go. We climbed a ramp onto a wide, two-level, roofless pedestrian walkway, inside of which sat the three tall stone pillars. I wondered about this ritual we were about to do. Samir saw the shape of the Washington Monument in the stone pillars. I saw giant phallic symbols rising into the sky. To me, the stone pillars represented something much deeper than the human struggle between good and evil. I believed evil came in many forms, one of them being the patriarchal oppression that has so suffocated women's spirits and voices throughout time. Pummeling the stone pillars represented for me the destruction of the forces that have domineered over women. When Muslim men use religion to oppress women, Islam is sullied, just as in any other religion characterized by such domination.

Months later I would listen to the journalist Bill Moyers interview the religion scholar Karen Armstrong. He asked her whether a religion could be true to its theological principles if it oppresses women. Absolutely not, she responded. I appreciated Bill Moyers asking this question as a man. And I so appreciated Karen Armstrong's blunt response.

I stared at these phallic symbols and flung my stones, one after another. "Allahu Akbar," I said under my breath with each stone I threw. God is great. A woman pushed me to get closer to the stone pillar. I didn't surrender to this obsession. Each toss of our stones was supposed to be a message that we could fight Satan, just as Abraham did. I didn't feel I had to muscle my way to a ringside seat.

With all our stones thrown, we gathered with our group to attempt a safe return.

MY NIECE'S MEMO TO ALLAH: A VERY GOOD GOD

MINA—A new day began.

It was the third day of hajj, Yawm al-Nahr, the Day of Sacrifice. Just as it's said that Allah tested Abraham's and Hajar's faith by sending Hajar and baby Ishmael into the desert, this day memorializes yet another supposed test of Abraham by God. It's said that Allah told Abraham to sacrifice his son, now grown. "Do as you have been commanded," Ishmael said. So Abraham put a blade to Ishmael's throat. At that point, an angel came to Abraham and ordered him to stop. Abraham had passed this test of faith and would be recorded in history as brave. Where is the courage in blind faith? Nothing is said about the whereabouts of Hajar—the courageous one, in my estimation—during this ordeal. We don't know whether she had died or was alive but not around.

Even this sacrifice was transformed into a battle of egos: Jews say it was Isaac who was about to be sacrificed, and Muslims say it was Ishmael. In Islam the day is Eid ul Adha, also known as Buk'reid, or "the Festival of the Sacrifice." It marks our second most important holiday after Eid ul Fitr, or "the Festival of the Fast Breaking," which follows the last day of the holy month of Ramadan. For both holidays, children get new clothes and the gift of cash called *eidie*. After Ishmael was saved, he and his father built the Ka'bah and beseeched God to make their descendants compliant.

I was one of the descendants of whom they spoke. They beseeched God to make *me* compliant. They called upon rituals and sought repentance. I had little of this spirit of surrender in me and never liked the holiday honoring their sacrifice at the altar of God, which always struck me as an odd test of faith. I wondered whether I would ever have so much faith in God as to sacrifice my child. Unlike Abraham, I would never put a knife to my son's throat, no matter how certain I was that God was commanding me to do just that. Today we feel that we don't have to challenge ourselves so deeply. We simply pay a fee to our travel agency to sacrifice lambs for our family and are told that the meat will go to pilgrims and the poor. The underlying principle goes much deeper, of course: to sacrifice what we love for the sake of God. To some, such sacrifice is also a way to act on their love of God.

To symbolize our willingness to sacrifice and symbolize our new birth into the faith, we got our hair clipped, trimmed, or shaved. My father and Samir got their hair shaved with an electric shaver. Back at the tent, men were giving themselves a smoother shave with razor blades. Safiyyah laughingly noted that my father's shave, with just a ring of hair around his bald head, didn't take long. But she took note of a few remaining strands on top. Samir wanted a closer shave. "Can I get one?" he asked. Outside the tents a young man, Saad Tasleem, stood over Samir with a razor. Samir liked these men. They were young and responsible. "They aren't lazy and stuff," he said. "I got my head shaved," Samir wrote in his journal, where he also drew a picture of himself.

Samir went outside the tent to feel the breeze on his bare skull. We were now released from most of the requirements of hajj—some of us more than others. Fourteen pilgrims were trampled to death on the King Khaled Bridge on their way to stone Satan.

Often over these days I had been told, "Fear Allah." Did I want to be motivated out of love of God, fear of God, or a combination of the two? To find my answer I looked around at what already motivated and inspired me. There was the day I spent some time with two women helping in our tent. The back flaps of the tent opened to the kitchen, where they slept and worked. They were robust, young women from Sudan, now living in Saudi Arabia; they had been hired to help during the pilgrimage. They slept on bedrolls on the floor, like us, beside a stove, with several toddlers and a baby who was a little older than Shibli. One of the women tugged at the long pants in Shibli's jumper and said something in Arabic. I didn't understand. I called a woman over who translated, "She would like a jumper with long pants."

This encounter evoked the traditional divide between the haves and the have-nots, as well as conflicting emotions in me. I felt imposed upon, but I wanted to share. I decided to share. After all, it was the hajj. I rifled through Shibli's clothes and found an outfit that had been a gift for Shibli from my former *Wall Street Journal* colleague Liz MacDonald, now a senior editor at *Forbes*. Raised a devout Catholic, she had once delivered heart medicine to Mother Teresa in Calcutta. When Mother Teresa had raised her hand toward Liz to bless her, Liz gave her a high-five, jolting the frail nun and sending the other nuns into a frenzy. Mother Teresa laughed. Liz

had gotten distracted and worried that she was going to hear another re-
cruiting speech to make her a Missionary of Charity nun. I knew Liz
would want me to give one of her outfits to a Sudanese baby in the middle
of hajj. I pulled out one with ABCs on it. The mother took the outfit and
another one I had. She didn't seem as grateful as I thought she would be.
It rekindled the conflicting feelings that such requests generate in me. Did
I want to help others out of love?

I found an answer about the inspiration that religion can be when I read a
memo that my niece had written to God in my notebook.

> From: Safiyyah
> To: Wonderful Allah
> Dedicated to Allah. Allah is the Best! Thanks to Allah we have
> food, water, shelter and many more things. Loving sweet Allah.
> Allah our 1 and only god
> Likes every one.
> Loveing
> A very good god
> HAPPY—Big Heart

It seemed that our Muslim society could be so much more proactive
about social services if we were motivated more by love of God than by
fear of God. One night our guide, Sheikh Alshareef, lectured about the
benefits of the social services that Christians provide in churches, like
hosting Alcoholics Anonymous meetings, giving out winter clothes, and
helping the needy at holidays. This was one of the first lectures my
mother fully appreciated. She said: "I'm a broad-minded Muslim. I believe
in service." I appreciated his philosophy too. The lecture caused me to reflect on my belief. As a Muslim, I wanted to encourage Muslims to engage
in more social service activities that would help people marked by taboo,
such as unwed mothers, alcoholics, drug addicts, and AIDS patients. My
mother and father had taught me from my earliest days to "serve humanity." I heard this mantra so often from my father that I would roll my eyes.
But at that moment, on the hajj, I realized that I needed to do more social
service work. And it also struck me how narrowly Muslim communities
often define their role in improving society. The solution in the most puritanical Muslim societies is simple: corporal punishment.
    Hearing Sheikh Alshareef's words and experiencing the kindness of my
fellow pilgrims made me realize that my life back home was a little out of

balance. I volunteered regularly at my niece's and nephew's schools, but beyond that I didn't feel as if I gave back to my community in Morgantown, whether Muslim or non-Muslim. I realized I wanted to be a more generous citizen, and I made a commitment to contribute to the social welfare of my community when I returned home.

I felt that only love, not fear, could change the tenor of the world, where the threat of violence loomed everywhere. *Arab News* reported on an alleged CIA warning about al-Qaeda attacks on the United States and Saudi Arabia as early as the end of hajj. More bad news trickled into Mina: another twenty-one people had been trampled to death under the King Khaled Bridge on their way back from stoning the devil. The Saudis reported that they would investigate the deaths and set up eight medical units to deal with stampedes. But faith can supplant good sense. CNN quoted a forty-five-year-old Sudanese pilgrim saying, "I am afraid to die, but this is a ritual that has to be performed."

What kind of public service was provided by these deaths? In 2001, 35 people died in a stampede. In 1998, 180 people died. The Saudis responded by turning to their Islamic scholars to bring rationality to the rituals. After another stampede killed pilgrims, Saudi Arabia's interior minister, Prince Naif, chairman of the Supreme Hajj Committee, was quoted on the front page of a newspaper calling on Islamic scholars to modify the narrow window of hours for stoning Satan. "I know that the scholars want . . . the pilgrims [to] perform the Hajj rituals following in the footsteps of the Prophet, but we have to look at this matter in the light of the present situation and problems," Prince Naif said.

If the Saudis could institute new interpretations of religion to deter stampedes, I wondered why they couldn't take the same position on so many more of the rules, restrictions, and rituals that didn't make sense.

CULMINATION

MECCA—As our final act in the pilgrimage, we were supposed to return once again to the Ka'bah to circumambulate it and perform Hajar's run between Safa and Marwah. We returned to Mecca to perform our sacred duty without a clue about the dangers that awaited us. Our bus dropped us off about a mile from the Ka'bah, the crowds were so big that night.

It felt as if every pilgrim was descending on Mecca from Mina that night, and they very well might have been. We walked on the road on which the Sheraton sat, passing the Kentucky Fried Chicken, the House of Donuts, and the Nigerian women peddling their rosary beads. As we neared, the crowd swelled in a way that we hadn't even experienced in Mina the day pilgrims had been trampled. The crowd pulsed with each step we took closer to the gates of the Sacred Mosque.

I remembered a time I'd just learned about. In the year 630, in this same place where crowds engulfed my family and me, the prophet Muhammad led ten thousand soldiers on a march to Mecca. The city surrendered without a fight. The once-despised former goatherder entered the city in triumph, went straight to the Ka'bah, and destroyed the idols of the 360 gods worshiped there. Later he reconstructed the shrine to Allah alone. Then, instead of taking vengeance on the city that had harassed and persecuted him for more than twenty years, the prophet declared a general amnesty. It was a new beginning, he said. "Allah be praised." Islamic history marked this juncture as the end of the period of Jahiliya, or ignorance. From Mecca, Islam spread quickly. Muhammad died in the year 632 after giving his sermon on Mount Arafat. Within a century the empire of his successors stretched from Spain to Afghanistan. It engulfed the armies of the Persians and the Byzantines and reached as far as the gates of Vienna.

I couldn't escape one thought. Just like any of the soldiers who accompanied the prophet Muhammad, we could die in Mecca. In our case, it would just take one mad scramble. My mother held tight to Safiyyah and Samir, their nimble bodies dwarfed in the crush of men and women who were not even aware of the young children near them. Strands of her hair slipped out from under her hijab and flirted daringly and defiantly with our eyes. Safiyyah felt an arm around her neck as pilgrims scrambled to move forward. "You don't think anything when you're about to die," she said later.

My father burrowed into the crowd like a mole in the subterranean layers of the earth. Shibli wasn't even in my arms. He was in the arms of a young man named Anthony Camadani, an American convert in our tour group with his two brothers. They had been a novel addition to our pilgrimage group of mostly immigrants. For starters, they didn't have the dark skin and black hair of most of our group. Second, they were cute, a thought I dared to entertain. Finally, they seemed to march to the beat of their own drummer—a refreshing change in this environment of cult ritual. Anthony's older brother, Daniel, threw his arm out to body-block for Safiyyah, whose svelte figure wasn't much defense against the crush of the pilgrims.

For his last visit to the Ka'bah, Shibli was in pink leopard pajamas—maybe not the best attire to tell him about when he turns twenty-one, but for the moment he was comfortable. He was eager and bright-eyed as the lights of the Sacred Mosque loomed in front of us. It was warm in the air, and Shibli was barefooted. He wiggled his toes and his face widened with a smile.

We had less than two hours to complete the final *tawaf*, or the circum-ambulation of the Ka'bah. To do this ritual is to complete the hajj. Gasping, we broke through the narrow entranceway, lucky not to have died. The crowds thinned out in the expanse of marble tile surrounding the walls of the mosque. We turned up the stairs and entered into this holy place. It was a madhouse. It would be impossible for us to join the scrum. We forced our way through the crowds and ascended to the third floor. The crush of people was enormous. It would be impossible there, too, to complete our rituals and catch our bus out of town. My father and I stood, staring at the crowd, confused about what to do.

"It's not safe for the children. You will be able to compensate," one of the young converts from our tour group proclaimed. "You can sacrifice a lamb to make up for it."

"Is that true?" I asked my father.

He shrugged his shoulders. He didn't know. It seemed impossible for us to accomplish our prescribed seven circumambulations in less than two hours. It had taken us all night to do it when we went there the first time. What to do? "Let's go," I suggested.

One person was clear about what we should do. "Forget about it," my mother said. "The children can't do it."

I thought to myself: *Oh no, the brakes again.* Samir was disappointed. "We can do it!" Samir pled. Safiyyah remained silent, staring at the crowd.

We tried to move forward but could barely move an inch. Completing our circumambulations seemed an impossibility. Maybe another floor? We slipped down a packed elevator to the second floor. It was even more crowded. We needed air. We returned to the roof.

"I don't think we can do it," I concluded sadly.

"Your father can go," my mother suggested.

"No, I will stay with all of you," he said. That was huge. My father made the sacrifice of staying with the family instead of completing his hajj ritual.

We settled for a picnic without the picnic food, sitting on the marble tile near a Saudi family. The husband gestured to Shibli. I smiled. He said something that I couldn't understand. I smiled politely, just a little frustrated that I couldn't figure out a solution for our final tawaf problem. His wife wanted to hold Shibli. I reluctantly passed him to her.

"I take baby?" the man said. I actually worried that he might be serious. I smiled and took Shibli back in my arms and drew him close to me.

Below us, men and women walked briskly on an interior hallway reenacting Hajar's run between Safa and Marwah. There was such an artificial feel to the mosque. It seemed remarkably fabricated. We peeked over the edge of the wall to look at the Ka'bah.

The Saudis had recently draped the Ka'bah in a new imported black silk cloth, called a *kiswa*, weighing about 1,500 pounds. They had done it while we stood in Arafat. It must have been much emptier that day with the pilgrims away. The Saudi Press Agency put a price tag of $5 million on the cloth. Saudi Arabia had set up a special factory thirty years before to design, manufacture, and tailor the cloth. That brought employment. But the price tag seemed like another unnecessary expense for a country with an impoverished class of people. The factory workers had decorated the upper half of the kiswa with a one-yard-wide strip of Qur'anic verses inscribed in silver threads painted gold, weighing about 250 pounds. Remembering the images of the poor under the King Khaled Bridge leading to our tent in Mina, such extravagance seemed outrageous to me.

It was time for us to leave. According to the rituals of Islam, we hadn't completed the hajj because my family, for safety reasons, didn't do the final circumambulation of the Ka'bah. I had to think about what *complete* means. There were actually two relevant concepts of *completeness* to think about—the physical and the spiritual. I didn't think about the former: though I felt uneasy about not physically completing the hajj, I knew I was doing the right thing for my niece, my nephew, and my son. To me, spiritual completion was more important. In the tent colony Sheikh Al-shareef's wife had told us: "The scholars say you will know your hajj is accepted if you go back from hajj a changed person." I knew that something meaningful had happened during our pilgrimage. I just hadn't had time to reflect and recognize any of the transformation that might have occurred within me. Not doing the circumambulation was important, but leaving safely from Mecca was even more important.

On the bus when we were leaving Mecca, I had a simple conversation with an Egyptian-born pilgrim who told me that the name Shibli means "*my* lion cub," not just "lion cub," as I had thought. This seemingly insignificant revelation was nevertheless quite important to me. I had chosen Shibli's name in part because I felt it captured a spirit of courage.

With Shibli in my arms, the storefronts of Mecca sweeping by our depart-ing tour bus, I realized that, indeed, my son was not separate from me. He was an extension of me. He was the physical manifestation of my courage. That courage had expressed itself on the hajj. I had come to Saudi Arabia thinking of myself as a criminal who needed to avoid detection and in the midst of a deep spiritual conflict over my son's conception. I left having nursed my son on the sacred ground of the Ka'bah, having liberated my-self with truth on the sands of Arafat, and having vowed to throw what-ever stones I could at the forces of darkness in the world. I was leaving complete.

Hajj pilgrims arriving at the airport in Jeddah, Saudi Arabia. (*Robert Azzi—Woodfin Camp*)

The Ka'bah, the Muslim Sacred Mosque at Mecca, Saudi Arabia. (*Robert Azzi—Woodfin Camp*)

Male and female pilgrims performing *tawaf* (circling the Ka'bah). *(Peter Sanders Photography, Ltd.)*

Running between Safa and Marwah to re-create Hajar's search for water. *(Peter Sanders Photography, Ltd.)*

Drinking water from the Well of Ishmael in Mecca. *(Peter Sanders Photography, Ltd.)*

Mount of Mercy in the valley of Arafat. *(Robert Azzi—Woodfin Camp)*

Shibli and me on Mount Arafat. *(Zafar Nomani)*

Paying at booths to arrange for an animal to be slaughtered on behalf of pilgrims. (*Peter Sanders Photography, Ltd.*)

Sheep waiting to be slaughtered for the Day of Sacrifice. (*Peter Sanders Photography, Ltd.*)

Pilgrims' tents in Mina, outside Mecca.
(*Peter Sanders Photography, Ltd.*)

Male and female pilgrims stoning the pillars in Mina.
(*Peter Sanders Photography, Ltd.*)

Male pilgrims having their heads shaved at the end of the hajj. *(Robert Azzi—Woodfin Camp)*

Safiyyah, my mother, and me in front of the prophet's mosque in Medina. *(Samir Nomani)*

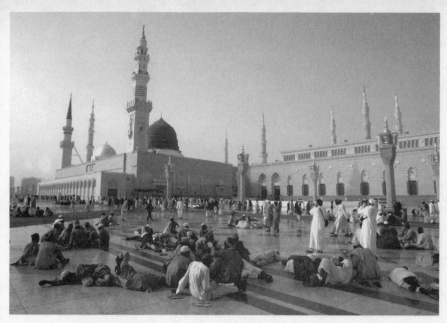

Muslim pilgrims gather outside the prophet Muhammad's mosque in Medina, January 24, 2004. (*AWAD AWAD/AFP/Getty Images*)

Entrance to the tomb of the prophet Muhammad in Medina. (*Peter Sanders Photography, Ltd.*)

(*Left to right*) Nabeelah Abdul-Ghafur, my mother, Shibli, Michael Muhammad Knight, me, and Saleemah Abdul-Ghafur before the march to the Morgantown mosque. (*Ron Rittenhouse*)

Prayer line before the march to the Morgantown mosque. Shibli in prostration. (*Ron Rittenhouse*)

PART FOUR

CONTINUING THE
PILGRIMAGE

February 2003

A DIFFERENT PATH

AMMAN, JORDAN—Royal Jordanian flight 2239 touched down at Queen Alia International Airport in Amman, the capital of Jordan, and my family and I glided off the plane, each one of us breathing a sigh of relief that we hadn't been arrested, detained, or harassed in Saudi Arabia. Our pilgrimage to Mecca and Medina was as complete as we could make it. We had one more important stop to make on our journey. We planned to visit the third holiest mosque in the Muslim world. It sat in Jerusalem. To get there, we were going to go through Jordan despite a U.S. State Department warning against travel in the region.

For me, Jordan was an important bridge between puritan Islam and the West. As we stepped through the tiled airport, my mother, Safiyyah, and I ripped off our hijabs. To some, that would have been an act of blasphemy. They believed that new *hajjahs,* or women pilgrims, should mark their rebirths by covering their hair. But if you believed the scholarly evidence that hijab isn't mandatory—as I did—then this wasn't an issue. Perhaps I also took off my hijab to defy the puritanical Islam in which we'd just immersed ourselves, with its rules, laws, and edicts so oppressive that we feared for our safety lest even a strand of our hair caught the glimmer of the sun. There was something fundamentally wrong with a society that was so totalitarian in its rule that a strand of hair could be a source of crime and punishment. The hijab had become a political symbol of Islam—a mandatory part of the uniform for claiming identity as a Muslim—and lost its real meaning of simple modesty. As a result, countless resources were spent by Muslim organizations, leaders, and activists on a single issue for women: covering their hair. It struck me as another way to divert attention away from more pressing issues, such as protecting women's rights to have a voice and provide leadership in our communities.

Muslim women had more freedoms in Jordan, including the right to choose whether to cover their hair or not. To understand why Muslim women had won more rights in Jordan than in Saudi Arabia, I remembered a book I'd read by the Moroccan feminist scholar Fatima Mernissi, *The Forgotten Queens of Islam*, about fifteen female rulers in the medieval world, women who reigned from Asia to North Africa as either monarchs in their own right or powerful consorts who acted as de facto heads of state. In Yemen two queens even had the honor of having their names announced in the mosque, a sign of their accepted power. Jordan has given the world models of strong Muslim queens in the modern day.

In 1978 a globe-trotting graduate of Princeton University married King Hussein of Jordan. I knew about Lisa Halaby, an American airline executive for Royal Jordanian, from my days covering the airline industry for the *Wall Street Journal*. She was the daughter of Pan American World Airways founder Najeeb Halaby, whose father was a Syrian Christian immigrant of Lebanese descent. Lisa converted to Islam upon her marriage, took the name Noor, meaning "light," and henceforth became Queen Noor. My honor as an author was that my book would be tucked in alphabetical order in the autobiography section at Barnes & Noble between Queen Noor's autobiography and Azhar Nafisi's *Reading Lolita in Tehran*.

Queen Noor had a profound influence on expanding the roles and responsibilities of Muslim women in Jordan. For example, women had become a part of the workforce. There were women working behind the counter at the Hotel Inter-Continental when we checked into our rooms. I never saw a single woman at work in Saudi Arabia except for the religious policewomen clad like ninjas. Queen Noor had succeeded as a strong-willed woman in Muslim society. That wouldn't be possible in most puritanical Muslim communities.

Amman is a cosmopolitan city with five Chinese restaurants and two French restaurants on its road map. The face of the new queen of Jordan, Queen Rania al-Abdullah, beams from a photograph tucked into a corner of the map, her mane of slick golden hair tumbling over her shoulders as she leans over a chair on which her husband, King Abdullah Bin al-Hussein, sits. Nowhere in Saudi Arabia did I see a single photograph of the queen of Saudi Arabia. The only image of a Saudi woman of royalty I ever saw was on a *Vogue* spread with a Saudi princess in a sultry dress.

The Star, Jordan's weekly newspaper, featured Queen Rania in a form-fitting black dress with tiny white buttons down her chest, her eyes glistening and her wide smile brightened by rosy red lipstick. I generally don't take notice of what a woman wears, but in the Muslim world women's dress is a window into ideology. Queen Rania looked nothing like the women I saw in Saudi Arabia. Did that make her less Muslim? I didn't think so. In the photo, a man is handing her the German Media Award for her work in improving the quality of the lives of women and children in Jordan and building tolerance. Queen Rania won the award jointly with Queen Sylvia of Sweden, the first women to win the prize.

Queen Rania was in the rank and file of the Muslim "generation X." Only five years younger than me, she was born in Kuwait to a notable Jordanian family of Palestinian origin. King Abdullah bin al-Hussein married her on June 10, 1993, when he was a prince. I had just turned twenty-eight and was recovering from my divorce. After the death of his father, King Hussein (Queen Noor's husband), the prince became king, and the princess became queen. After that, Queen Rania moved easily between villages in rural Jordan, empowering women's cooperatives, and five-star hotels in Paris, coordinating microfinancing projects, her sleek hair her hijab. She was a woman to whom I could relate. Why, I wondered, did Saudi Arabia and other puritanical societies have to order women to cover their hair? It should be a personal matter left to choice, I felt. Nowhere in the Qur'an is it explicitly stated that women have to cover their hair or abide by a specific dress code. There are two references to modesty but they don't even mention scarves. "O Prophet! Tell your wives and your daughters and the women of the believers to draw their cloaks over their bodies [when outdoors]. That is most fitting that they could be known as such [i.e. decent and chaste] and not molested," says Verse 59 in Chapter 33 of the Qur'an. And, then, in Verses 30–31 of Chapter 24 of the Qur'an, it says: "Say to the believing men that they should lower their gaze and guard their gaze. And say to the believing women that they should lower their gaze and guard their modesty; that they should not display their beauty and adornments except what (must ordinarily) appear thereof; that they should draw their veils over their bosoms and not display their beauty except to their husbands, their fathers . . . (and a list of exceptions)." The strict customs of veiling and secluding women in separate areas came three or four generations after the prophet's death when Muslims adopted traditions practiced by Greek Christians of Byzantium. Islamic scholars have widely differed in their interpretations of how hijab should be applied.

What became obvious to me in Jordan was that the Saudi brand of Islam doesn't define Islam, and—most importantly—it doesn't have to define Muslim society. For one thing, it felt freer and more festive in

Jordan. The Amman Marriott advertised a program that merged America's Valentine's Day with Islam's Eid, the Muslim holiday of sacrifice that had just ended: "Romancing Valentine's into Eid." And when our guide took us to meet his newly arrived aunt from the hajj, sparkling red lights framed the front door, like the kind used on Christmas trees in the West. A bright sign covered the front door. It said, HAJJ MABROOR (an "accepted hajj"). This wasn't just a reflection of Jordan being entrapped by Western globalization; it was clear that Saudi Arabia was equally susceptible to the presence of multinational corporate branding. Rather, Jordan was a society that seemed to allow for more plurality of expression.

But it was also clear that women didn't have their full freedoms. When we went to Kentucky Fried Chicken in Amman, an image of Colonel Sanders greeted us, and the hostess welcomed us with friendly conversation and toys for the children, but she expressed frustration with the restrictions her family imposed on her, invoking Islamic law. "I'm trapped," she said.

Women were raising their voices against inequities and injustices with legal appeals, and there were still wrongs to be corrected. Women were campaigning for seats in Parliament. They were in the *Jordan Times* criticizing as too small the six-seat quota the government had just assigned women. They were led by inspiring human rights activists such as Asma Khader and May Abul Samen. Asma Khader was a lawyer and founder of the Sisterhood Is Global Institute, an organization that created literacy and legal assistance programs for Jordanian women. Elected to the Permanent Arab Court as counsel on violence against women, she had become a leading advocate of a campaign to outlaw honor killing. About twenty years earlier, a mother's plea drew her into the cause of defending women. The mother had come to her office for help. Her husband had murdered their fifteen-year-old daughter, who had become pregnant after being raped. Her husband had received only a six-month sentence because he claimed that he had killed his daughter for the family's honor. The shock: the mother revealed that the rapist was the father. The court believed his denial.

Advances had hardly brought full justice to women. Asma Khader had most recently been defending a young Muslim woman in Yemen, Layla Radman Aiesh, whom a religious court had sentenced to be stoned to death for *zina*, or illegal sex, while the man with whom she had had sex was sentenced to one and a half years in prison. In my research on Jordan, I discovered a gloomy picture of mothers' rights in Jordan at Iexplore.com, a travel website. "Child custody decisions are made in religious courts; it is difficult for an American woman, even a Muslim, to obtain custody of her children in an Islamic court unless she agrees to stay in Jordan. Husbands/fathers may deny permission to travel to their wives and children, regardless of religion or nationality." The site included a link to the U.S. State Department's website and a phone number for reporting child abductions.

Nevertheless, Jordan served as a model of a Muslim society coexisting peacefully with other faiths and histories. My family and I climbed by car to a place called Mount Nebo, where Christians say Moses was buried. I stood where the pope had reveled not long before in the beauty of the Holy Land below. Of course, on the other side of the Jordan River, toward Palestine and Israel, there sat a building where the Muslims say Moses was buried. We also wandered through the ancient ruins of Petra, where one of the Indiana Jones movies, *The Last Crusade*, immortalized the pre-Islamic civilization that flourished under the Romans. In a sign of her blessing, Queen Noor had visited the set.

THE PROPHET AND A MIRAGE

Glory be to He
Who carried His servant by night,
from the Holy Mosque
to the Furthest Mosque,
the precincts of which
We have blessed
so that We might show him
some of our signs.
Surely He is the All-Hearing,
the All-Seeing.

"Al-Isra" (The Night Journey),
Qur'an 17:1

JERUSALEM—My name came from the seventeenth chapter of the Qur'an, "Al-Isra," or "The Night Journey." *Asra* means "to journey." The passage tells the story of a magical journey that the prophet Muhammad supposedly made from Mecca to Jerusalem, flying through the skies on a mystical winged creature. Yusuf Ali, a renowned Qur'anic translator from modern-day Pakistan whose lyrical translations I'd grown up reading, called the tale of the *miraj* (mirage) "a fitting prelude to the journey of the human soul in its religious growth in life." I'd always had a dream to live out the journey that was my name.

The story goes that in the ninth year of Muhammad's mission, about 620 A.D., when he was forty-nine, the prophet was sleeping in the holy mosque in Mecca one night when the angel Jibril, or Gabriel, stirred him awake. He led the prophet to the edge of the mosque. In one version of the story, Jibril had the prophet Muhammad prostrate and took out his heart, removing a blood clot in it and saying, "That was the part of Satan in thee." Then Jibril washed the prophet's heart in a golden basin filled with *zamzam*, the holy water of Hajar, healing it and putting it back in the prophet's body. Jibril brought the prophet a white-winged creature called *al-baraq*, "an animal white and long, larger than a donkey but smaller than a mule." "Each stride stretched as far as the eye could see," the story says. The prophet mounted the strange creature, and it flew him to a mosque in Jerusalem. This was an important symbolic destination, for Jerusalem was the focus of the first qiblah, or direction of prayer, until a dramatic moment in Medina. Dissuaded by his conflicts with local Jewish and Christian leaders, the prophet Muhammad turned the qiblah to Mecca to distinguish Islam from Christianity and Judaism.

When the prophet arrived in Jerusalem, the story says, he entered the mosque and prayed two rak'at (prostrations). It's said that Abraham, Moses, Jesus, and other prophets prayed behind him. When Muhammad came out, Jibril brought him a vessel of wine and a vessel of milk. He took the milk, and Jibril said, "You have chosen the true religion." It's said that Jibril then took the prophet to Heaven and opened the gates of Heaven. "Who are you?" a voice asked.

"Jibril," he answered.

"Who is with you?"

"Muhammad."

"Has he been sent for?"

"He has indeed been sent for."

The gates opened. "Lo! We saw Adam," the prophet told his friends. Adam welcomed him and prayed for his good. Then Jibril took the prophet Muhammad from the second heaven to the seventh heaven, and on each level he saw another prophet—Jesus, known as Isa in Arabic; John the Baptist, or Yahya; Joseph, or Yusuf; Enoch, or Idris; Aaron, or Harun; Abraham, or Ibrahim; and, finally, God, or Allah. After tough ne-gotiations with God, the prophet whittled the mandatory number of prayers a Muslim must say every day from fifty to five. As the story goes, Moses persuaded the prophet to keep going back to God to reduce the number, round after round, even urging him to go one more time to reduce the five mandated daily prayers, but the prophet balked at doing this, noting, "I returned to my Lord until I felt ashamed before Him."

Jerusalem held great importance for Muhammad, as the last prophet in the Abrahamic tradition. It was near Jerusalem in Hebron where Abraham supposedly lived with Sarah and Hajar before Abraham took Hajar to the desert. As the patriarch of Judaism, Christianity, and Islam and the pioneer of monotheism, Abraham made this city important to all three faiths. The Old City measured less than a square mile. All three religions revered this city. In the Muslim world, Jerusalem was al-Quds, or "the Holy."

After the prophet's death in 632, the Muslim ummah spread beyond Mecca and Medina, but it was always fraught with the tension between worldly ambitions and spiritual aims. The first caliph was Abu Bakr (632–34), a companion of the prophet; he united Arab tribal chiefs in the lands around Mecca and Medina during his two-year reign. Taking power in 634, the second caliph, Umar ibn al-Khattab (634–44), conquered much of the Persian Empire. In 638 he conquered Jerusalem and subsequently controlled Syria, Palestine, Egypt, and the North African coast. When Umar entered Jerusalem in 634 he guaranteed freedom of worship for all faiths. Finding ruins covering the rock from which the prophet ascended to the heavens, he cleared the rubble and built a mosque nearby and called it al-Aqsa, meaning "the Distant Place." The area was known as al-Haram al-Sharif, or "the Noble Sanctuary," but also was referred to as al-Aqsa. In November 644, a Persian prisoner-of-war stabbed Umar in the mosque of Medina. Six of the prophet's companions elected a man by the name of Uthman ibn Affan (644–56) to succeed him that year. Uthman belonged to a family called Umayyad, and his armies seized lands from Tripoli in what is today Libya to the eastern parts of the Persian Empire. Discontented Arab soldiers assassinated him in his home in 656. A fierce five-year civil war broke out between allies of the prophet's son-in-law, Ali, and a man by the name of Muawiyah, the son of the prophet's old enemy, Abu Sufyan. Muawiyah won and power became entrenched in the Umayyad family during his rule, from 661 to 680.

During the period of rule by Umayyad leaders that followed, Umayyad caliph Abd al-Malik started work on enclosing the rock of the prophet's ascension into a grand mosque and expanding al-Aqsa into a more ornate mosque. His son, al-Walid, apparently completed it in 705 as a symbol of

Islam's permanence in Jerusalem, a city with a Christian majority. It has remained essentially unchanged for over thirteen centuries. The mosque is called the Dome of the Rock, and its glittering golden dome is Jerusalem's most visible icon.

In a twist on Abraham's story of sacrifice, Jews believe that Abraham was ready to sacrifice his son Isaac, not Ishmael, on God's command. For Jews, the site of that sacrifice is the same place where Muhammad ascended to the heavens, and Jews call it the Temple Mount. Jews venerate it as the place where Solomon built a temple that was destroyed and then rebuilt by Zerubbabel. The puritans of Judaism and Christianity believe that God will miraculously destroy the mosque through earthquakes or other means, leading to the building of a third temple, where, they say, Jesus, or the Messiah, will rise again. Meanwhile, the Muslims have claimed it for themselves. And we have all fought for control over this parcel of land.

After control over the Dome of Rock switched through many hands, a Muslim Kurd leader in the twelfth century named Saladin reclaimed the Dome of the Rock Mosque, called by some in the Muslim world "the jewel of the signet ring of Islam."

Over the centuries Jerusalem fell under the jurisdiction of many rulers, from the Canaanites to the ancient Hebrew kings, the Islamic caliphate, the Christian crusaders, the Ottomans, the British, and now the Israeli government. What has transpired in the region is one of the most dramatic and endless conflicts in the world. The roots of the conflict lie in the historic claim to the land that sits between the eastern shores of the Mediterranean Sea and the Jordan River.

For the past one hundred years Palestinians—who are mostly Muslim but also Christian—have faced colonization, expulsion, and military occupation in the land of their father, Abraham. For the Jewish people of Israel, the land represents a return to the land of their father after centuries of persecution.

Key to the struggle is the place where we were going to try to pray together as a family—the holy sanctuary where the prophet is said to have ascended to the heavens. To this day the prophet's ascension makes the mosque the third-holiest mosque in the Islamic world, after the mosques in Mecca and Medina.

Modern-day politics over the creation of the state of Israel transformed the experience of the al-Aqsa mosque for Muslims. In 1948 Jewish settlers proclaimed the state of Israel. British troops left the region, and fighting broke out between Israel and its Arab neighbors, ending in October 1949 with some 700,000 Palestinians fleeing or being driven from what had been British-mandated Palestine. Israel annexed large tracts of land and gained control of Jerusalem in the west; Jordan took over control of Jerusalem in the east, where the Dome of the Rock and al-Aqsa sit. In June 1967 war broke out between Israel and its neighbors. The Arabs were convinced that they could defeat the small nation of Israel with their collective might. But in six days Israel defeated the Arab armies, giving the conflict its name—the Six-Day War. In its victory, Israel grabbed wide stretches of territory from Jordan, Syria, and Egypt. In a symbolically devastating move, Israel seized control of East Jerusalem and the ancient Old City from the Jordanian troops who had controlled it. For the first time, the al-Aqsa mosque and the Dome of the Rock were under the control of Israel.

Visiting this mosque used to be a part of the Muslim pilgrimage after the hajj, but few Muslims ventured there after the latest chapter of trouble in 2000. In September of that year, Ariel Sharon, at the time an Israeli opposition leader, ventured into the courtyard of this holy space in Jerusalem's Old City while Muslims had gathered inside for the Friday midday prayer, considered the holiest in the week's prayers. Sharon arrived with a small army of soldiers and police "to see what happens . . . in the holiest place of the Jewish people." Men assembled for prayer threw shoes at him and yelled, "Allahu Akbar" (God is great) and "Murderer, get out," as he moved from the compound's west side to the east and back again. Palestinians responded with an intifada, or uprising, which became the Palestinian resistance to Israeli military forces. Israel clamped down on Muslim tourists in Jerusalem.

A wall-sized photo of Jerusalem with the trademark golden dome hangs on a wall in a restaurant on High Street in downtown Morgantown, but I had never understood its political, historical, and religious symbolism. As a Muslim, I had only started learning about the noble sanctuary, with its holy mosques, when I was preparing for our pilgrimage there. Before we left America, I had started to read Bruce Feiler's best-selling book, *Abraham: A Journey to the Heart of Three Faiths*, but I couldn't pretend to make sense of the claims and counterclaims to this land. I couldn't even keep the name of the mosque with the golden dome straight. The Mosque of the Golden Dome? The Golden Dome Mosque? *The Dome of the Rock. The Dome of the Rock. The Dome of the Rock.* I tried to memorize the name but kept getting caught in mental tongue twisters.

One result of Arab-Israeli tensions was that my generation had been robbed of a real appreciation of this historical place. When Israel took control of the Dome of the Rock during the 1967 war—two years after I was born—it fell off the map of pilgrimage for my generation in part because of the disdain by many Muslims, particularly the puritanical, for Israel, coupled with the difficulties that Muslims experienced traveling there.

Despite this history of deterrence, we set off for Jerusalem from Amman to pray on the third leg of the pilgrimage. As we left, the question for us was simple: could we get inside Jerusalem?

TRAVEL WARNINGS AND WAR CLOUDS

ON THE ROAD TO JERUSALEM—Just before we left the United States, the State Department had issued a travel warning. "The Department of State warns U.S. citizens to defer travel to Israel, the West Bank, and Gaza." It cited "numerous civilian deaths and injuries, including to some American tourists, students, and residents. The potential for further terrorist acts remains high." It warned Americans to stay away from restaurants, cafés, shopping malls, pedestrian zones, public buses, and bus stops in Israel. To cover any place it might have missed, the warning added "other crowded areas and venues" to the list of off-limits sites. It mentioned Jerusalem as a place in which travelers should be particularly careful.

CNN had reported that the White House raised the terrorist alert level because of the hajj and that the U.S. was pulling diplomats out of Jordan and Israel. To get to al-Aqsa, we would be going straight into the heart of the warning—Jerusalem—and we were going to be at public tourist sites and staying at a public hotel.

"Let's not go," my mother said quickly.

"No, it'll be fine," I said with no authority at all.

As we were leaving our hotel in Amman, war clouds loomed over the Middle East. Journalists from John Burns of the *New York Times* to the *Washington Post*'s David Ignatius and ABC's Diane Sawyer had come through Amman to cover the impending war in Iraq. In Saudi Arabia the *Arab News* published a story with the dateline "Occupied Jerusalem" and the headline "War in Two Weeks."

News reports said Israel had issued an emergency call-up for reservists to man Patriot antimissile systems. And tensions remained between

Israelis and Palestinians. The *Jordan Times* reported: "The Israeli occupation army pressed on with the relentless arrests of suspected Palestinian resistance activists, netting more than two dozen in swoops on Palestinians celebrating Eid ul Adha, the festival of sacrifice."

The front page showed four young Palestinian boys on a donkey cart in the Jabalia refugee camp in the Gaza Strip, watching the funeral procession of two young Palestinian men who had been killed trying to infiltrate the Dugit Jewish settlement. While we were in Saudi Arabia, the *Saudi Gazette* had included a column from the government-appointed imam of the mosque in Mecca, Sheikh Dr. Abdul Rahman al-Sudais, that stoked the fires of hatred for Israel and Zionism, the movement that created Israel:

> The Muslim ummah is being targeted by overt and covert conspiracies of the enemies such as the Zionist whose evil actions in Palestine, and [their onslaught] against the al-Aqsa Mosque, the third holiest place of Islam, are [clear] indications of arrogance and terrorism. Allah who protects the Holy Ka'bah and the two Harams [the sacred places of Mecca's and Medina's mosques] shall also protect al-Aqsa Mosque from evil machinations.

Without fear, my father had worked hard to make our trip happen. He had downloaded pages of information from the Internet, circling the number of kilometers from Amman to the King Hussein Bridge (thirty-five miles), underlining the biblical sites we could see (the Mount of Olives and Mount Zion), and scribbling down the e-mail addresses of guides. The two sides couldn't even agree on what the King Hussein Bridge was called. Israel rejected Jordan's name and called it Allenby Bridge, for General Edmund Allenby, who seized Jerusalem for the British in 1917 after attacking the ruling Ottomans in Palestine.

With politics creating such a divide between people, the question that plagued us remained the same: were we going to be able to get into Israel? We read and reread the visa and travel advisories. "Israel does not require advance visa issuance for U.S. citizens traveling on tourist passports at any crossing point." I found a travelers' dispatch. It advised us that we could get a visa at the border. It was easy enough for us to get a bus across the Jordanian border, but when we hit a certain point it was obvious that Israel controlled its borders like a hawk. Inside, the border police studied our documents carefully. My father leaned in toward the immigration officer, a young woman, and asked her not to stamp our passports with an entry into Israel. Certain Muslim countries deny entry to anyone who has traveled to

Israel because they don't recognize its existence. "Please stamp another piece of paper," he kept telling the immigration officer. She didn't respond. In the line beside us, the immigration officers were now interrogating a busload of Palestinian pilgrims just returning from the hajj. I watched a young female soldier turn an elderly woman pilgrim's Qur'an upside down, leafing through its pages to see if anything would fall out. I tried not to let my horror show on my face as I watched. It seemed like such an indignity. "Please stamp another piece of paper," my father said again.

"Dad!" I yelled, tugging him away before he annoyed the officer so much that we would get our foreheads stamped with a big good-bye.

As it turned out, Safiyyah, Samir, and Shibli were our special visa into Israel. With the children, we didn't appear threatening. Finally, we were released and stepped into Israel.

It was parched with occasional trees, but mostly vast tracts of barren land filled the horizon. Immediately, I thought: *they can't figure out a way to share all this land?* I had thought the territory was jam-packed with settlements, cities, and other development. Quite the contrary—Israel was mostly empty. Granted, there were legitimate issues about irrigation and water supplies, but for all of the billions of dollars spent on defense, surely there was a way to make all of this land viable and share it. I felt downright defeated as I contemplated these vast stretches of territory, and no sense of celebration on entering this land. The modern-day cycle of violence, tragedy, and hatred, quite simply, made it depressing to be there.

Still, we had the small issue of a pilgrimage to make. A man told us our guide was waiting at the end of the road right past the security checkpoint. The Israeli military didn't let him wait for us at the border crossing. We took a taxi a few hundred yards down the road, where we stopped beside another car. From it emerged a Palestinian man whom our Jordanian travel agency had hired as our guide. My father apologized for our delay. Our guide had been waiting for hours on the Israeli side for our arrival. He wasn't bothered.

"What's a few hours?" he said. "I have been waiting three years for business."

The scenes remained depressing. Barbed wire atop chain-link fences lined the road as we drove toward Jerusalem. The barbed-wire fencing surrounded Palestinian refugee camps. Everywhere we looked on the road to Jerusalem we saw Palestinians who seemed to be living in prison camps. Towering over the settlements in the hills were well-constructed townhouses facing Jerusalem. "They're the settlements," the guide told us. "*Settlements?*" I asked. "The Jewish settlements?" I was stunned. They looked like a tract of suburban living plucked right out of any town in the United States. The scene changed as we climbed into the Mount of Olives, a flattened series of ridges just east of Jerusalem and a place where Jesus walked. We stopped at a convenience store, where my father videotaped the scene. The guide coaxed the cashier to "smile." "There is nothing to smile about," the Palestinian cashier responded flatly. I saw the truth in what he said.

I wanted to go to Hebron, where Abraham supposedly had lived with Sarah and Hajar before taking Hajar into the desert of Mecca. "Impossible," the guide said. The U.S. travel warning said that sections of the West Bank and Gaza had been declared "closed military zones." It added: "The government of Israel may deny entry at Ben Gurion Airport or at a land border to persons it believes might travel to 'closed' areas in the West Bank or Gaza or to persons the Israeli authorities believe may sympathize with the Palestinian cause and are seeking to meet with Palestinian officials."

The warning didn't bode well for us.

THE DOME OF THE ROCK

JERUSALEM—The prophet Muhammad arrived in Jerusalem on his winged creature. We pulled up to Lion's Gate on the northeast corner of the Noble Sanctuary in our guide's car. In a grim reminder of the politics of the day, a funeral procession passed by us at just that moment.

The sacred ground covers over thirty-five acres of fountains, gardens, buildings, and domes, or almost nine times the size of the grounds of the U.S. Capitol in Washington, D.C. Al-Aqsa sits at its southernmost end, and the famous Dome of the Rock sits in its center. The entire area is considered a mosque and makes up almost one-sixth of the walled city of Jerusalem.

We went through one of the many security checkpoints Israel has set up within its borders, including in Jerusalem, to stop terrorist attacks such as suicide bombings. Israeli soldiers checked our bags. Safiyyah felt sad about all of the guns that soldiers around her were carrying. Samir stared at the soldier. He thought to himself, *He has a gun*. It was not a scene he normally saw in West Virginia. Samir was scared. What was he scared of?

He had heard about how Muslims and Jews were fighting over the land. He knew Palestinians who had lost their homes when the British remapped Palestine after World War II to create the state of Israel. "It's wrong that the Palestinians lost their homes," Samir said. He had watched CNN reports on Palestinian suicide bombers. He had thought, *They're so desperate they're killing themselves.*

In Saudi Arabia the headlines in the *Riyadh Daily* had given me a reality check about the politics of the country. "Crown Prince Slams Campaign Against Islam," one headline read. Crown Prince Abdullah bin Abdulaziz, the deputy premier and commander of the National Guard, said that there was a campaign against the prophet Muhammad. "In fact, only some spiteful people and the Zionist quarters don't like the prophet," the newspaper quoted him as saying. Reading the article, I was reminded of the September 11 conspiracy theorists in Pakistan. "The Yahudi did it" ran like a mantra through conversations as Pakistanis from elementary school children to physicians blamed the attacks on New York and Washington on the Jews. *Yahudi* is an Arabic word for any Jewish person born after the fall of the ancient kingdom of Judah. Those born before that time, dating back to Moses, are Bani Isra'il.

This was the kind of rhetoric that had led to the painful clashes that erupted into centuries of world wars. I had learned the concept of "enemy image making" in graduate school, and it seemed to me that was what we were doing in every corner of the world. We created enemies of each other. "This is a cycle of hatred that just isn't ending," I told my family. I turned to my mother. "Imagine if Sarah had just told Abraham, 'Go take a hike,' and raised Isaac with Hajar raising Ishmael. The world might have been a very different place."

We slipped through the stone archway and stepped into a vast open space where the golden-domed mosque loomed in front of us. In the sixteenth century, Suleiman the Magnificent, a ruler of the Ottoman Empire, had commissioned artisans to inscribe the Qur'anic verse "Ya Seen" across the top in spectacular tile work. Testifying to the prophet Muhammad and his teachings, it is considered the heart of the Qur'an.

> *By the wise Qur'an.*
> *Surely you are among those sent on a straight path.*
> *A revelation of the Mighty, the Compassionate.*
> *That you might warn a people whose fathers were never warned, so they*
>   *are heedless.*
>
>   Qur'an 36:2–6

We entered through the front door of the Dome of the Rock, together as a family, as we had done in Mecca. It was a wide, round mosque with an expanse of space. There was no separate women's entrance, and no gender segregation inside. There were no women's hours to see the rock, which we immediately saw in front of us, from where the prophet supposedly ascended to the heavens. It was massive and solid. It was quite an impressive image in the middle of this building.

To avoid any distractions by human imagery, the interior included no figurative art. In *Islam: A Short History*, the scholar Karen Armstrong evokes an Islamic concept of monotheism called *tawhid*, or the oneness of God, to analyze the symbolism of the Dome of the Rock. She writes:

The dome itself, which would become so characteristic of Muslim architecture, is a towering symbol of the spiritual ascent to heaven to which all believers aspire, but it also reflects the perfect balance of tawhid. Its exterior, which reaches towards the infinity of the sky, is a perfect replica of its internal dimensions. It illustrates the way in which the human and the divine, the inner and the outer worlds complement one another as two halves of a single whole.

Our guide took us down a short stairwell to a lower-level view of the rock. Four elderly Palestinian men stood beside it in conversation. "Assalaam alaykum," said one with a friendly smile. "Walaikum as salaam," we responded. The interaction was ordinary but memorable to me. I realized that it felt welcoming to be greeted so cordially and respectfully both here and in Mecca. I wasn't used to such kindness. Back home the community had become so hard-core that men often wouldn't even greet women who walked right beside them at the mosque—when women even bothered to go.

We said a prayer; my father told me it was considered blessed to give a special prayer when entering a mosque. When we finished, we went across the courtyard to the second mosque, al-Aqsa. It was built on the site of the original timber mosque constructed at the time of Umar. Our guide told us that the mosque overflowed with worshipers on Fridays for the holy prayer. I found a quiet spot in the massive expanse of empty space inside the mosque and enjoyed a prayer alone.

It was surreal to realize that thousands have died because of this spot on earth. Around me were showcases that had bullets on display, shot from Israeli weapons at Palestinians.

With another prayer finished, we returned to the Dome of the Rock. Inside we prayed the sunset prayer together. My mother, Safiyyah, and I prayed in a section with a row of about six other women, about ten yards behind a row of men and boys that included the guide and my father and nephew. Although we were behind the men, the trapezoid form of the mosque kept our position from feeling subordinate. Nothing felt separate. I lay Shibli beside my feet. It was so beautiful to pray there together.

After we finished our prayer, I absorbed the presence of this holy place once more. Then we stepped outside into darkness. The sky drew my eyes upward. The clouds swept past the moon—full and wide—in a glorious testimony to the divine. The rush of the clouds darkened the moonlight for just a moment, as if it were the divine breath itself that sent the clouds whirling through the night sky. I was spellbound.

"Look!" I exclaimed to my family, urging their eyes upward. They turned their eyes to the heavens and absorbed this celestial moment.

"Al-hamdulillah!" my father exclaimed. "Praise be to God!"

"Neat," said Samir.

"Cool," said Safiyyah.

"Let's go," said my mother, always so practical. "It's too cold. Shibli will get sick."

The golden dome was an awesome sight, stretching 66 feet across the Noble Rock and rising to a point more than 115 feet above it. We stood in the quiet and empty courtyard, staring at the heavens as we stood near the rock between earth and heaven. We had ascended into our own mystical journey. It took me where most roads lead—home.

PART FIVE

BRINGING THE PILGRIMAGE HOME

March 2003 to October 2003

A TRIUMPHANT HOMECOMING

Whereas the hadj is a culmination for most pilgrims, it felt more like a starting point to me.

> Michael Wolfe, *The Hadj:*
> *An American's Pilgrimage to*
> *Mecca* (1993)

MORGANTOWN—When we returned to Morgantown, winter was thick in the air. I returned to my parents' house with Shibli and threw myself into the momentum of my life as a new author and new mother. Family called from all over the world to congratulate us on completing the hajj. Friends came over to drink from the *zamzam*—the holy water that we brought back.

My first book, *Tantrika: Traveling the Road of Divine Love,* hit the bookstores as spring flowers started to bloom. I was proud of the honesty with which I had written about my life's journey, including Shibli's conception, but I knew that speaking truthfully could be costly. A fatwa had been issued against Salman Rushdie in 1989 for his fictional writing. How would Muslims respond to the truths I wrote about sexuality, worship, and identity? I braced myself with the only weapon I had: knowledge. For three days I sat in the last row of a lecture hall on the campus of George Mason University in an unusual course: fatwa school. My teacher was none other than my guide on the hajj, the young Canadian Muslim Sheikh Muhammad Alshareef.

I donned the same hijab I wore in Mecca. Camped out beside me, Shibli played in his Evenflo Portable Fun UltraSaucer, absorbing vicariously the systematic process of Islamic jurisprudence, or *fiqh* (rhymes with *pick*). Fiqh refers to the bodies of law collected through interpretation by scholars into the different schools of jurisprudence that have defined the Muslim world for centuries. By the end of the weekend I had learned the first lesson in liberating myself from fear of reprisal. For any fatwa issued to condemn me, I could find a fatwa to affirm me. Sheikh Alshareef called it fatwa shopping. As I drove away from the weekend seminar, I saw one of my fellow students in full black nikab, looking like a Saudi Arabian woman. I read the bumper sticker on her car: "If you think education is expensive, try ignorance." Even though she and I probably ascribed to different interpretations of Islamic law, we at least agreed on that one point.

My first review arrived with the delivery of our local Morgantown newspaper, the *Dominion Post*. "Nomani bares soul in beautifully written tale," it said. This recognition meant so much to me, even if it came off the pages of my local paper, because I had received so many cultural messages marginalizing the value of women's voices and so still struggled with insecurities.

In New York I sat down for an interview with a reporter for *India Abroad*, a weekly magazine I'd grown up reading. I had never felt fully understood by the people of my culture, but the reporter, Aseem Chhabra, pierced my skepticism. He related to my identity struggle. The next week I stared at the result of his work. "Divine Love," the headline read next to a full-page photo of me smiling adoringly at Shibli as he stood on my lap on the banks of Manhattan's East River, gazing eagerly off camera where my parents were making funny faces at their grandson. *India Abroad* represented to me the rigid expectations of my traditional society. But here on its pages my honesty was celebrated.

The first test of how far I was willing to go with my newfound clarity came when an e-mail shot around the world and into my in-box: Amina Lawal, a Nigerian woman who'd had a daughter out of wedlock, would be stoned to death if she didn't win a pending appeal. I stared at the warning, looked at my blossoming son, and thought: *There but for the grace of God go I.* I felt as if I had to take a stand. Her possible death forced me to catapult into action because of the life that had sprung from within me. To even my own surprise, I found the answers and liberation in Islam itself. I found brothers in Islam who didn't act as judge, jury, and executioner of Muslim women. In Rhode Island, Omid Safi, a Colgate University professor and the editor of *Progressive Muslims: On Justice, Gender, and Pluralism*, rejected the criminalization of mothers for *zina*. "Islam has a strong tradition of humane judicial reasoning, or *ijtihad*, which is used to mediate questions of law," he told me.

"*Ijti*-what?" I asked. I'd never heard this Arabic word before.

"Ijtihad," he said. "It is based on *istihsan*, equity, and *istihsal*, the needs of the community." Dr. Safi grew up in Iran, where his mother became his icon for an ethic of compassion. Like her, he was dedicated to devoting himself to love and service to others.

Alan Godlas, the professor of religious studies at the University of Georgia, exchanged e-mails back and forth with me to explain how post-colonial Muslims, fueled by Saudi Arabia's financing of Wahhabi doctrine, were turning to puritan Islam to set themselves apart from the West.

That political and theological analysis made sense to me. In the name of religion, puritanical Muslims were unleashing a vigilante force on women. Yet they were imposing man's law, not God's law. I had entered Saudi Arabia as a criminal in the eyes of Saudi sharia, guilty of zina. I prayed in the holy epicenter of Islam. And God did *not* send a thunderbolt down to earth. I survived. Islam survived. The Ka'bah survived. Karen Armstrong explains in *Islam: A Short History* that by the fourteenth century community leaders known as the *ulama* "liked to believe that these laws had been in place from the very beginning of Islamic history. Thus, while some Sufis, such as Rumi, were beginning to glimpse new horizons, many of the ulama believed that nothing ever changed. Hence, they were content to believe that the 'gates of *ijtihad* were closed.'" Reeling from invasions, "the *ulama* had transformed the pluralism of the Quran into a hard communalism, which saw other traditions as irrelevant relics of the past. Non-Muslims were forbidden now to visit the holy cities of Mecca and Medina, and it became a capital offense to make insulting remarks about the Prophet Muhammad."

I knew what I wanted to say. Muslim societies would do well to heed modern thinkers who would have us love the soul—both the mother's and the child's—without loving the sin. The sting of shame and stigma is emotionally as deadly as a physical stoning. I chose, however, to liberate my son and myself with the truth, because I didn't want him to inherit a legacy of shame. To punish mothers seemed to me to be not only cruel but a violation of the Universal Declaration of Human Rights, which states in article 25, section 2: "Motherhood and childhood are entitled to special care and assistance. All children, whether born in or out of wedlock, shall enjoy the same social protection." The shame and stigma so prevalent in traditional Muslim societies also struck me as a rejection of the value Islam put on motherhood.

It's said that a man asked the prophet, "Who among the people is the most worthy of my good companionship?"

"Your mother," the prophet replied.

"Then who?"

"Then your mother."

"Then who?"

"Then your mother."

"Then who?"

"Then your father."

After 9/11, Muslims and Islamic associations distanced themselves from terrorism perpetrated in the name of Islam. I argued that with the trial of Amina Lawal, Muslims should renounce the harsh punishment of women for such intimate "crimes" as zina and join with human rights and women's rights groups, such as Amnesty International, that are defending women indicted for zina. I saw that I was not alone as a Muslim opposed to the criminalization of women's sexuality. A Muslim woman in Malaysia, Zaitun "Toni" Kasim, helped found Sisters in Islam to fight sexism that criminalizes women's bodies. In Pakistan, a team of lawyers led by two sisters, Asma Jehangir and Hina Jilani, defended women charged with zina. And a Muslim woman named Ayesha Imam, a lawyer and activist, was leading the defense of Amina Lawal in Nigeria through her group called Baobab, which was also defending other women in Nigeria. I remembered the term *baobab* from the story *The Little Prince*. The Little Prince, who comes from an imaginary planet, describes the baobab as a tree with roots that could overrun his small planet if untamed. Ayesha Imam and other Muslim activists chose Baobab as the name of their women's rights organization because in Africa the baobab is a tree of strength.

I discovered that in Morocco, where the police expect doctors to report single mothers giving birth at hospitals, a group called Solidarité Féminine runs a job-training shelter in Casablanca for single mothers. Morocco's King Mohammad VI decorated the founder, Aicha Ech-Chenna, despite fundamentalist Muslim protests that she and her shelter encourage prostitution. She was trying to help children of single mothers obtain identification cards, which they need for schooling but can't easily get because they're dismissed as illegitimate. I was amazed at this activist's drive. She was also pushing for DNA testing to establish the paternity of these children.

It was clear to me that if Muslim society is to mature, we must move away from punitive responses to the realities of this world. Why not support mothers instead of stoning them? It seemed particularly important to reconsider the issue considering that there is no consensus on the issue of punishing zina. The clerics who condemn mothers to death simply sub-scribe to the most puritanical interpretation of Islamic law from among the four diverse—and often conflicting—schools of jurisprudence that have survived into modern day in the Sunni sect of Islam. And there is no reference to stoning in the Qur'an as a punishment for zina.

It seemed obvious to me that zina laws are not a humane, fair, or judi-cious response to social realities. Because it is difficult to find the four wit-nesses required by the Qur'an to prosecute zina, the men involved are usually released. Pregnancy, however, is telling, so pregnant mothers are imprisoned even though majority sharia opinion concludes that preg-nancy is circumstantial evidence.

I knew the sad truth. It is women—rarely men—who become targets of these punitive sex laws. Increasingly, women charged with zina have ended up in jail, often with their babies, to face the sentencing of death or lashings while the father goes free. The man Amina Lawal named as her baby's father denied paternity and was released. In 2002 the man whom Zafran Bibi, a Pakistani woman, charged with raping and impregnating her also denied the allegation and was released. She was sentenced to death by stoning—a ruling that I read about while I was in Pakistan, care-fully guarding the secret of my own pregnancy. Her sentence was later overturned. But every day there are untold numbers of Muslim women who abort their pregnancies, dump their babies in rubbish piles, or se-cretly abandon their children so that they won't have to face the conse-quences of having a child out of wedlock.

I felt good about one thing. I wasn't going to be lashed. I was in West Virginia. There I had the support of loving parents and was shielded by the progressive laws of a country where religion and state are constitu-tionally separated and where consensual adult sexual behavior has largely been decriminalized. Every morning, like mothers everywhere, I playfully nuzzled my infant son, drawing a smile from him. Shibli was my joy, but in Saudi Arabia and other Islamic countries run by sharia, he was more than that. Shibli was proof that I, an unmarried Muslim woman, was guilty of zina.

Many cultures and religions inveigh against premarital sex and adultery and consider these activities sinful. But I'd learned that much of Western society and moderate Islam have long concluded that humanity and God are best served by separating the sin from the sinner. I could accept the re-buke of those who believed that what I did was wrong. I understood that there would be consequences for my son and me as I struggled as a single mother to raise him. Although my judgments about men may not always have been great, my son would always be a joy to me. I certainly didn't believe I deserved to be jailed or lashed for what I had done.

In countries such as Saudi Arabia, home to Mecca and the spiritual center of Islam, tradition has perpetuated these sex laws, which were propagated in part by the strict Wahhabi movement. In countries such as Nigeria and Pakistan, Islamists have implemented or strengthened these laws and enforced them rigorously. Muslim societies that punish women for alleged crimes of the body contradict the fundamental principles of forgiveness, privacy, and motherhood in Islam.

A mother is considered a *jihadi*, or holy warrior, by virtue of conceiving her baby. It's said that if she dies in childbirth, she becomes a *shaheed*, or martyr. As a society, Muslims should support those who reject the criminalization of sexuality and are risking their lives to protect mothers and their babies.

Pending her appeal, Amina Lawal wasn't incarcerated, but she was imprisoned by judgment and condemnation. When I saw the photo of her daughter, Wasila, I thought of my son, crawling by my feet as I wrote. I thought about how in my native India, all elders are "aunties" and "uncles." Amina Lawal would be "Amina Auntie" to my infant son. I pulled out a T-shirt on which I wrote, "Free Amina Auntie." Shibli and Wasila are children of the same generation from different worlds in homes defined by the same religion. Both deserve futures with their mothers.

As the prophet Muhammad said, "Heaven lies at the feet of your mother."

I sat at my laptop and wrote an essay defending the rights of women such as Amina Lawal and me to live without punishment for alleged crimes of our bodies. It was a huge departure from my usual silence in simply being a spectator but not a commentator. The *Washington Post* published the essay in its Outlook section, "She Shouldn't Be Stoned. None of Us Should."

When my article was published, I was ready for a fatwa condemning my challenge to puritanical Muslim jurists' interpretation of sharia. The newspaper ran the image of Shibli and me on the banks of the East River juxtaposed dramatically with a grim picture of Amina Lawal sitting alone on a bench facing a panel of male judges. To my surprise, I received about fifty e-mails from Muslims expressing their support.

A Muslim associate producer from the Canadian Broadcasting Corporation's *Newsworld Today* program, Layal El Abdallah, e-mailed me an invitation: "We would like you on air!" I asked her why she was interested in the issue. Like me, Layal came from a supportive Muslim family, and her parents encouraged her to be a journalist. She had been following the story of Amina Lawal and trying to find someone who would speak about the issue so that she could broadcast it. It started becoming clear to me that the world was desperate to hear the voices of Muslim moderates. A CNN producer called me from Washington. "We'd like to have you on air on tomorrow's morning show." I had always been a print journalist. Even when the *Wall Street Journal* teamed up with CNBC to put *Journal* reporters on air, I resisted. But confronted with the dearth of Muslim voices speaking out against actions like the sharia sentence against Amina Lawal, I wanted to bring a human face of reason to the issue. Battling stage fright, I proceeded, remembering the encouragement of the many Muslim men and women who had written expressing their support. The interviewer was a CNN veteran, Leon Harris, an African American anchor. I listened to his disembodied voice through my earpiece and argued my case for mothers such as Amina Lawal.

"So what happens to the men in these cases? What happens—for instance, what happened to the father of your son?"

I stared into the camera. How could I answer that question delicately? "Well, he couldn't kind of commit and get beyond the cultural and personal divide that is not even about religion and personal issues sometimes—"

Harris interrupted me. "What? He abandoned you?"

I stammered. I realized how uncomfortable I was admitting the truth. "Well—you want to—you don't want to be so—you know, you hope that." I paused to collect my thoughts. "I'm not answering you clearly, because—I'm struggling with it, aren't I?"

"It sounds like it," said Harris. "Why is that so hard to say, if that's what happened?" How could I explain the decades of programming telling me that a woman must be chaste? "I know, it is hard to say because ultimately, this is the psychological dimension of this issue, so even when you're not physically stoned for this kind of 'crime,' you are struggling with a lot of the . . . shame. I mean . . . it wasn't my plan to *not* have a father for my child when I [chose] . . . to be a mother. And so, this is the personal issue that we all have to deal with when you decide to take on tradition."

Harris seemed to understand. "That's amazing. And that's how pained you feel about it, and you're not in jail right now somewhere in Nigeria, as Amina Lawal is. We'll have to continue to follow her story, her case, and perhaps we can talk about all of this again with you down the road. Fascinating."

In that awkward TV moment, I realized how far I still had to travel to accept the fact of my single motherhood. But I knew that I had a personal ethic to which I was trying to remain true. I had committed myself to it on my judgment day in Arafat.

MY VOICE CLAIMED

NEW YORK TO LOS ANGELES—That summer I crisscrossed the United States, often with Shibli and my family, on my book tour, a concept both foreign and daunting to me. As with the hajj, it felt right to have Shibli with me. He had helped propel me into a new identity. I was a Muslim woman who had searched for her voice. I had expressed it powerfully as a journalist and writer. But in the world of journalism, typically, we work very hard reporting and writing a story. When it gets published, we typically just move on to the next story. We hardly go out there to speak about our messages. It was now my time to express my thoughts through my literal voice. And I was afraid.

"I am overcoming my own fears as I stand before you," I told the audience that had packed a gallery not far from the Williamsburg Bridge across the East River from Manhattan. I immediately broke the first rule in public speaking by admitting my inadequacies. I was marked by the silence of women's voices in the political debates that ignited the men's section at the dinner parties my parents hosted. Even when my mother sent me into the men's room to quiet my father's boisterous position statements ("In its foreign policy, America must practice the democratic ideals it preaches!"), I would always whisper. My senior year at Morgantown High School, I ran away from the microphone when our local radio station WVAQ wanted to interview me for the Superstars Day sports competition I'd organized for the school. A childhood in the kitchen had taught me to remain in the background. I rarely claimed my public voice.

As an adult, I connected with the American feminist writer Susan Faludi's words when she wrote, "I am woman. Hear me whisper," in a *New York Times Magazine* essay. She had been my colleague at the *Wall Street Journal*, and she wasn't kidding. At an orientation session for new reporters, Susan literally whispered answers to questions. Her front-page story about the human casualties of leveraged buyouts had been so powerful that it had won a Pulitzer Prize, but in a conference room in the *Journal* headquarters in lower Manhattan we could barely hear her. I could write powerfully, but I was afraid to speak publicly. With my news stories, I had helped send a young scam artist to jail by revealing his fraudulent business practices at Braniff Airlines; I had exposed the secrets of alleged price fixing at the country's major airlines; I had challenged CEOs, CFOs, and corporate raiders like the fiery Carl Icahn. But I had never really set forth my views about . . . well . . . anything. Now all of a sudden I had to speak, and my words were causing a stir. I was shocked. So many people stay silent because they believe, as I did, that others know more than they do or that their thoughts are irrelevant. When I realized that my own thoughts were not only relevant but well informed, it occurred to me that many people have important things to say but remain silent because of their fears. Going on the hajj clarified for me that I had to speak out and challenge some of the antiquated traditions in the Muslim world. When we go against conventional wisdom, we just have to be braced for criticism. People attack those who challenge the status quo.

Indeed, when senior editor Deborah Caldwell posted an interview with me on a religion and spirituality e-magazine called Beliefnet, I received slurs I didn't even know existed. One Muslim called me a *munafiq* for revealing the truth of Shibli's conception. "What does that mean?" I asked my mother, stirring masala, or spices, into that night's dinner.

"Hypocrite," she told me. "They're stupid."

Even with my mother's support, the allegation stunned me. It was an Arabic word I had never learned. To motivate me, my parents had taught me not punitive judgments but rather positive ideals. Instead of teaching me about the vice of being a munafiq, they taught me the virtue of being truthful. It was ironic to me that I was being called a hypocrite just as I was baring my soul for the first time in public.

I returned to the computer a few hours later and scrolled through the comments with my father. Another Muslim called me *mushrik*. "What is *that?*" I asked my father. Reading the comment board had become one of my father's favorite postretirement hobbies, much to the annoyance of my mother, who hadn't yet figured out how to send an e-mail. "Someone who commits *shirk*," he told me, looking up from his screen. He always seemed to have the screen tilted so that he would have to point his chin up a little and read through the bottom of his bifocals.

"*Shirk?*" I'd never heard this word before. I checked on the Internet. *Shirk* means worshiping other gods besides Allah, or God. It is pitched by the puritanical scholars as one of the "major sins," along with murder and zina, giving me a batting average of .666, according to some Muslims. I knew my study of Hinduism and Buddhism would not go over well with all Muslims. After all, in Nepal, I had bowed my head to Kali, the dark goddess of Hinduism who slays evil. On the roads through the Himalayas in India, I had meditated to Durga, the fierce Hindu goddess who sits on a tiger. In trying to understand the other faiths, I didn't feel as if I was violating the one into which I was born. I was motivated by the Qur'anic injunction to think and to learn. My experience with Buddhism and Hinduism didn't leave me believing in many gods and goddesses. Instead, the way I looked at it, they had simply deified the many expressions of one divine force. Islam, after all, has ninety-nine names for Allah: merciful, compassionate, wrathful, loving, and all-knowing, among them. With visual representations considered un-Islamic, Muslim artists over the centuries have illustrated these manifestations in intricate calligraphies, sometimes depicting the words in the shape of animals. The great Buddhist and Hindu scholars told me that their religions believed too in a singular, wider divine force.

My father rejected these slurs that Muslims, and people of all religions, threw upon others to dismiss dissent. "They're crazy," he said as he tapped the keyboard. I agreed with him. And I had to look no farther than the pocket-sized guidebook I'd gotten in Mecca on the rules of the hajj to understand where this ideology was rooted: Saudi Arabia's puritanism. "The Things That Nullify *Iman*," a section of the book began. *Iman*, meaning faith, had become a politically correct concept that Muslim puritans used to distinguish themselves from other Muslims. Of course, they took ownership of the term to define themselves as the true people of faith. To not have faith was to engage in *kufr*, nonbelief.

The number one violation was to commit shirk.

Number two was speaking to God through "intermediaries" like a saint.

Giving polytheists, or *mushrikeen*, any legitimacy was the number three violation. I didn't begrudge Hindus their faith. I respected it. I was guilty as charged.

Number four was giving credibility to "some guidance other than the Prophet's guidance." Of course, that meant we had to accept their interpretation of the prophet's "guidance." Violations included believing that secular laws and systems are better than sharia; "the Islamic system is not suitable for application in the twentieth century"; "the Islamic system is the cause of backwardness of Muslims"; Islam "is only a relation between a man and his Lord, and does not have any relations with other aspects of life"; "the judgments of Allah in enforcing the punishments prescribed by Allah, such as cutting off the hand of a thief, or stoning an adulterer is not suitable in this day and age"; ". . . it is permissible to rule by a law other than what Allah has revealed in Islamic transactions or matters of criminal justice and similar affairs." Adhering to any other law besides sharia, according to the guidebook, had horrible consequences: "Anyone who regards as permissible something that Allah has made impermissible, such as adultery, drinking alcohol, or usury (loaning money at interest), such a person has become an unbeliever according to the consensus of all Muslims." It was an exaggeration to say this conclusion would be the "consensus of all Muslims." The Saudi religious establishment was discounting debate among Islamic scholars and jurists over these points and finding a convenient way to legitimize itself.

Violations number five through ten were all variations of the same theme: allegiance to the Qur'an, sharia, and the traditions of the prophet, without any discussion about which traditions were credibly recorded and which were not. To me, these were the rules of blind faith, and I didn't buy them.

Neither did my father. He was creating aliases under which he defended me in these public message boards. A friend of mine later aptly described the comments on these message boards as akin to graffiti on bathroom doors. Fortunately for me, my father was right in there with the best of the punks.

ANGER

NEW YORK—As I stood at the Astor Place Barnes & Noble talking about the injustice of killing mothers for alleged sexual crimes, a young Muslim man stood up and politely, but still directly, admonished me for being a criminal in the eyes of Islamic law. He wanted to hear from my father, as if only my father could defend me. I argued that my position was legitimate even in the eyes of Islamic law, but bowing to this idea of my father as the patriarchal spokesman of my life, I handed the podium over to him. I didn't even realize that my mother, seated in the audience, could just as easily answer the question.

A consummate professor, my father turned the book reading into a lecture. I cringed as he went on and on about the teachings of compassion in Islam, the value of mothers in the Muslim world, and his faith in me. I tuned him out. Then, all of a sudden, I heard thunderous applause for my father. He had just told the audience, "I love my daughter. I love my grandson."

Torn between gratitude for my father and a resentment I couldn't fully understand, I took the microphone back. That night I exploded. We were standing in front of Curry In A Hurry, a restaurant in the middle of Little India in midtown Manhattan. "Why do you have to talk so much?" I screamed. "Why do you have to go on and on and on? I *hate* you!" My anger was clearly disproportionate to the situation, but I was so consumed by it that I couldn't clearly see what I was really angry about.

The tension between my father and me kept escalating through the summer. It peaked with a showdown late one Sunday morning in July as I prepared to speak at Washington and Lee University about women in the Middle East. I hadn't slept all night, trying to get my thoughts organized.

My mother insisted that my father drive so that I wouldn't risk falling asleep at the wheel. That time I just didn't have the energy to battle my father's monologues. "No! I don't want him to come!" My mother insisted. I stormed out to find my father sitting in the driver's seat with the door locked. "No!" I yelled. "You're not going!" It was a moment out of a teen soap opera. "I'm not going to go with *him!*" I shouted with a piercing gaze. My mother returned my dagger eyes. "Don't talk about your father that way."

"You're a mother now! You have to stay alive," she said, clutching Shibli in her arms and turning quickly back into the house. I strode wildly behind her. She was in the living room beside a round marble coffee table. She looked frightened at my rage. "I know! You think I don't know I'm a *mother?*" I yelled. To my mother, I had suddenly become a stranger she did not recognize. "You hate us!" she screamed.

My father finally relented, getting out of the driver's seat, and I tumbled inside, slamming the door behind me. It was not a Kodak moment in maturity. Rolling through the gentle hills of southern West Virginia into Virginia in our family's Suzuki SUV, I realized the significance of what I had just done. I had refused my mahram—my sanctioned male chaperone. I was a woman driving alone to give voice to my thought. In my fatigue, my deeply rooted resentment at the patriarchal forces in my culture expressed itself in anger. In a way, I did hate my family. I didn't hate them literally but figuratively, for all that I felt I had inherited from them—the silence, passivity, subordination, and compromise expected of Muslim women. What I couldn't see clearly then was how they had freed me from my own inheritance.

As I drove I wondered about the depths to which I had sunk in claiming my independence from my family. I ended up with so much self-loathing because I couldn't seem to find a mature way to free myself from the programming I wanted to escape. I understood why I had turned to writing. Growing up, I hadn't felt as if I could express myself fully. My parents were struggling with their own feelings of displacement. They loved me deeply, I knew, but each time I went to them with feelings of alienation or discomfort, I couldn't fully express myself. It was a timeless and universal issue, but the particular dynamics of my Muslim culture exacerbated the problem for me. When I couldn't find a sympathetic ear in my family, I turned to my writing. I could at least talk to myself through confessional writing. My notebook never told me to stop my inner dialogue because it was leading me off the "straight path." My journal never said, "Huh?"—the guttural sound that was the Urdu equivalent of "What?" and my father's response to my views when he was absorbed in his own thoughts and didn't pay close attention to my every word as I impatiently expected.

My journal never expected me to censor myself.

If I paused, I focused on the strengths that I had gained by being a daughter of my family—opportunity, voice, strength, and certainty. But in my moments of frustration I felt defined by my weaknesses more than by my strengths. In this moment of self-criticism, I wanted to disappear. My mother was right. I had become crazed. That was what this tension of the spirit did to my soul. As I drove toward Lexington, Virginia, I felt like a fraud going before the Washington and Lee alumni to talk about women in Islam. That night on campus I tucked myself into bed with hopes that I would awaken with a new fortitude.

Instead, in the morning I felt depressed and couldn't pull myself out of bed. The clock ticked and ticked. I turned on the radio just to be able to drown in noise. The conversation revolved around the NBA basketball star Kobe Bryant and the charge of rape against him. This is what happens, the fundamentalists argued, when a woman doesn't live by *hudud*— the boundaries of a moral code. If she'd had a male chaperone with her, she wouldn't have been raped, the argument went. I would always get tripped up on that logic because it was literally true. I concluded that freedom of choice carries risks but allows for great benefits to society and human beings.

The phone rang. The organizers were politely wondering if I would be joining the conference soon. I couldn't dawdle any longer. I pulled out my notebook. I was clear now about my outline. I gave it a name: "*Hudud*: Sacred Boundaries—the Intersection of the Political and the Personal." I knew what I would talk about. My life. Born in India. New Jersey. West Virginia. Journalism. Hudud: the physical: harem, clothes, veils, mobility; sharia laws: Pakistan, Malaysia, Morocco, Saudi Arabia, Nigeria; zina; hudud breakers: Amina Lawal, Ayesha Khan, Asma Jehangir, Toni Kasim, my mother.

I ventured over to the conference and sat in the back of the amphitheater where the group was listening to a scholarly discussion by a professor. I felt intimidated. The organizer introduced me, and I stood before the jury. I spoke and spoke and spoke. My life story unfolded easily, and the larger points about the cultural limitations and greater possibilities for women in Islam spilled out easily. The audience was attentive and then volleyed question after question at me. They recognized me as legitimate, and I felt credible. At that moment of triumph I missed my father. But I knew I had had to venture out alone to know my own personal strength. On the drive home I bought Shibli a battery-operated hopping rabbit, Beat Box Bunny, that delighted everyone at home, where the fallout from my departure was dissipated in my safe return.

ACCEPTANCE

CHICAGO—We had gone on the pilgrimage with the Islamic Society of North America, and I had been pleasantly surprised at how affirming they were of women's voices. The society was holding its annual convention, and I threw caution to the wind and registered to go with my father and Shibli. I had never gone to one of its conventions before, afraid of not fitting into the scene.

I went, in part, as a journalist. I wanted to know the answers to questions such as Islamic law's ruling on Amina Lawal. She was facing a trial date for her appeal, and I wanted to know whether I was intellectually or theologically flawed in defending her. My opportunity to find out would come during a news conference at which the society's top woman leader was going to speak. She was Ingrid Mattson, a scholar and convert. She was the one my father had told me about, selling me on the idea that the society was open-minded.

I took a seat in the second row, where it dawned on me: I wanted a seat at the table in the Muslim world. I didn't want to remain in the shadows, as I had done not only in the Muslim world but even in my life as a journalist for the *Wall Street Journal*. I remembered who I was. I was the reporter who never asked questions at press conferences. I was the reporter who dared ask a question of the Dalai Lama only when I felt that the stakes were too high to allow demonization, intolerance, and suspicion to have the final say. I did something awesome for me: I stood up and moved to a seat in the front row between a public radio journalist and another reporter. And I raised my hand. "The Islam that the West is seeing is an Islam that sentences a woman in Nigeria to death for having a child out of wedlock. Is this an accurate depiction of Islam?"

The secretary general of the society stood at the podium first and shook his head in understanding at my question. A native of Kashmir, India, his name was Sayyid Saeed. He was a congenial-looking man with a head of thick graying hair and a salt-and-pepper beard. He said that verdict reflected only one interpretation of sharia, and he personally rejected this interpretation, as did many scholars. With that, he gave the podium to Dr. Mattson. She was one of the scholars who rejected the verdict of stoning, and she gave a poised and powerful defense of her position, arguing that Islam's teachings on compassion and tolerance override any perceived sexual crime. I felt so affirmed that my religion didn't condemn a woman such as myself and that in America I had the intellectual freedom to find that out. More than in any Muslim country, this was where I could be a fully realized Muslim woman and mother. "We should not judge women," Dr. Mattson said.

As the press conference closed I felt so buoyant about being a Muslim woman in America. This was where I could live with my head held up high, as my religion told me I deserved to do. Dr. Mattson held her head up high, silver thread glittering off the hijab she wore. I held my head up high, my black hair just touching my shoulders.

THE LEGACY OF THE PROPHET AND HIS DAUGHTER

MORGANTOWN TO ATLANTA—In Chicago, I met Michael Wolfe for the first time in person. His father had been Jewish, his mother Christian, and he exemplified plurality. He was warm, engaging, and normal, relating to me. He interacted with respect, but ease. I appreciated this after growing up in an immigrant community in which men kept an uncomfortable distance from women. I asked him if he could recommend anyone in the American Ismaili Muslim community with whom I could talk. He suggested that I talk to a friend of his, a California businessman by the name of Amir Kanji. In Chicago I also met Alex Kronemer, Michael Wolfe's co-partner in film making. I had an easy conversation there with both of them about identity, faith, and the challenge of Islam in our time. Amir Kanji, I discovered, was also easy to talk with about these things.

A distributor of halal food (ritually fit according to Muslim law) out of Northern California, he gushed with excitement about expecting his first grandchild soon. His eldest daughter, Shari, had married a white American Christian man and was now pregnant. I was stunned at how easily he talked about it. A woman marrying outside Islam was anathema in the Muslim culture in which I had grown up. The Qur'an (5:5) allows Muslim men to marry Jewish and Christian women: "This day are all things good and pure made lawful to you. . . . Lawful to you in marriage are not only chaste women who are believers, but chaste women among the People of the Book, revealed before your time." But non-Muslim men are forbidden to Muslim women. The Qur'an (2:221) says: "Nor marry your girls to unbelievers until they believe. A man slave who believes is better than an unbeliever."

Accepting this edict, I had turned away from a man who loved me deeply. He was a white Lutheran man from Iowa. I broke his heart—and only later did I realize my own as well—to marry a Pakistani Muslim man who was right for me in every way but substance. I had never met Muslim parents who accepted their child's marriage to a non-Muslim. "That's okay with you?" I asked. "We are Ismaili Muslims," Amir told me. "Living in North America, Ismaili Muslims, like other Muslims, are learning to live with the reality of inter-faith marriages.

"*Ismaili?*" I answered.

I had heard this word mentioned only a few times in my life, but I had no idea what it meant. Ismaili Muslims are a part of a minority sect of Islam that doesn't subscribe to some of the mainstream interpretations of Islam. For example, they don't impose a ban on Muslim women marrying non-Muslim men. It was a right I had never even allowed myself to think could be possible. A while later I read that a religious body in Turkey reversed the ban. Only then did I realize just how much suffering this edict had brought to my life. I was an American. For years I had denied it. But there was no denying my cultural affinity for the values that I had learned growing up in West Virginia.

In trying to understand this school of Islam that expanded my relationship with Islam, I got a history lesson in politics in the early days of Islam. What I learned was that, as undignified as it sounded, leadership after the death of the prophet came down to a chick fight. In one corner was Aisha, the young and favored wife of the prophet. Her father was Abu Bakr, an elder statesman. After the prophet died in 632, his companions sided with Aisha and elected Abu Bakr the first caliph of Islam. The line of caliphs starting with Abu Bakr had the allegiance of people who later became the roots of the Sunni sect, the majority of believers in Islam. Ismaili Muslims, on the other hand, as part of the minority sect of Islam called Shi'a, believe that the first rightful caliph of Islam should have been Ali, the husband of Fatima, the prophet Muhammad's favorite daughter. In a land where blood relations are sacred, Ali, a cousin of Muhammad's and the son of a man named Abu Talib, was the prophet's closest male relative. Although he was young and inexperienced, the Shi'a believe the prophet designated Ali to be his political and spiritual successor and that Fatima lobbied for him. The followers call themselves Shiah i-Ali, "the Partisans of Ali," or simply Shi'a. Although she lost the battle over succession, Fatima became the namesake for the Fatimid line of caliphs, which traces its lineage to her and Ali.

Ali Shariati was an Iranian sociologist and thinker who may have been killed by the Shah of Iran's secret police before the Shah's overthrow in 1979. A Shi'a, as are most Iranians, he lectured to his students in Iran about Fatima. "The visage of Fatima," he said, "the visage of the woman who existed, who spoke, who lived, who played a role in the mosque, in society, in home training her children, in her family's social struggles and in Islam; a woman whose role should be made clear in all its dimensions to the present generation (not only to Muslims, but to any human being, man or woman, who has human feelings, who believes in human values, and who is faithful to real freedom) should be accepted as the best and most effective role to be imitated by the present generation."

The prophet called his daughter Fatima one of the four greatest women in the world. When she was a girl, she walked hand in hand with her father as he brought social revolution to Mecca and Medina. When his enemies once dumped dust on his head from a balcony, she wiped the dust from his face. In exile with her family in the desert outside Mecca, she comforted her elderly mother, Khadijah, and her older sisters. Her sisters lived important but ordinary lives with their husbands and families, while she had a place in Islamic history. When it came time to marry, she continued her life as a revolutionary, accepting the hand of Ali, a man who could offer her nothing but love and, literally, the proceeds from the sale of a shield. Taking little dowry money from her husband gave her a place in history. When I was married, I took this "dowry Fatima," which evokes an independence and self-sufficiency in which I believe. As a mother, Fatima dispatched her sons, Hussein and Hasan, to learn how to lead under the tutelage of their father and grandfather. On her lap and in her home, she raised her daughter, Zainab, to become a revolutionary and a pioneer. All the while, her father believed in her. He called her "the woman among the women in the world," establishing her as a role model. Over the protests of companions of the prophet, she fought for the right to own a small farm called Fadak. Despite their differences, even Aisha, the prophet's wife, said, "I never saw anyone higher than Fatima, except her father, the prophet."

As part of the Shi'a sect, Ismaili Muslims live in about twenty-five countries, mainly in South and Central Asia, Africa, and the Middle East, as well as in North America and Western Europe. The achievements of the Fatimid dynasty dominate accounts of the early period of Ismaili history, roughly from the beginnings of Islam through the eleventh century. The Fatimid dynasty created a state that stimulated the development of art, science, and trade in the Mediterranean Near East over two centuries. Its center was Cairo, founded by the Fatimid as their capital. Following the Fatimid period, the Ismaili Muslims' geographical center shifted from Egypt to Syria and Persia. After Alamut, their center in Persia, fell to Mongol conquerors in the thirteenth century, Ismailis lived for several centuries in dispersed communities, mainly in Persia and Central Asia but also in Syria, India, and elsewhere. The leader of the Ismaili Muslims is called their imam, and they call him "His Highness the Aga Khan." In the 1830s the Shah of Persia granted Aga Hassanaly Shah, the forty-sixth Ismaili imam, the honorary hereditary title of Aga Khan. In 1843 the first Aga Khan left Persia for India, which already had a large Ismaili community.

The most recent imam, the forty-ninth hereditary imam of the Ismaili Muslims, became the leader on July 11, 1957, at the age of twenty, succeeding his grandfather, Sir Sultan Mahomed Shah Aga Khan. Ismailis have a history uncommon among Muslims. This imam, Karim Aga Khan, was born on December 13, 1936, in Geneva, spent his early childhood in Nairobi, Kenya, and then attended Le Rosey School in Switzerland for nine years. He graduated from Harvard University in 1959 with honors and a BA degree in Islamic history. The Aga Khan would set forth a philosophy that established Islam as "a thinking, spiritual faith," teaching compassion and tolerance and upholding the dignity of women and men.

He saw it as his mandate to "safeguard the individual's right to personal intellectual search and to give practical expression to the ethical vision of society that the Islamic message inspires." I connected to the way the Ismaili world wanted to express Islam.

In a speech as long ago as 1976 in Karachi, the Aga Khan argued that the prophet Muhammad's example should inspire Muslims to create "a truly modern and dynamic society, without affecting the fundamental concepts of Islam." That was music to my ears, but blasphemy to the ears of puritanical Muslim leaders who prospered teaching intolerance, hate, and the status quo. Still, Ismailis were breaking new ground. In Canada, the home to many in the community's diaspora, an Ismaili woman, Yasmin Ratansi, was poised to become the first Muslim woman elected to Parliament.

One of the few contexts in which I'd heard about the Aga Khan was through an elite university he endowed in Karachi, aptly called Aga Khan University, or AKU to alumni. I went there to interview a physician, a woman, about depression and suicide in the country. Wandering through the campus, I wondered whether I had somehow found myself on the campus of UCLA in southern California. "Dude, where you headed?" I heard a young man in a baseball cap ask a friend. When I turned around, I saw a gaggle of hip young Pakistani Muslim men and women who had a Western air about them, just like me, only ten years younger.

During our hunt to find my friend Danny, I had taken his wife, Mariane, to one of the Aga Khan Hospital clinics for pregnancy-related blood tests. When I discovered I was pregnant, I went to the clinic hesitantly for my own testing. In the back of my mind, I wondered whether Pakistani law enforcement officials would seize the results of my blood tests and use them as evidence of my pregnancy, and thus zina. After I had left Pakistan and was in Paris, at the start of my second trimester, I asked Shibli's biological father over and over again to pick up those test results, but he broke promise after promise and never got them. I always wondered if he just never wanted to be associated with this proof of my pregnancy.

As I talked to Amir Kanji, the Ismaili halal food wholesaler, I wondered if my results were still sitting in the hospital files. But Amir rejected the criminalization of zina and the other expressions of puritanical Islam, particularly in the West. "We need to create an American Islam versus being just Muslims in America." He invited me to Atlanta for a meeting of the Ismaili Health Professionals Conference. I boarded a flight to Atlanta with Shibli, not knowing what to expect.

A PRINCESS IN ISLAM

ATLANTA—My education in the freedoms possible for Muslim women in modern-day Islamic society began at the check-in counter at the Atlanta Westin. The two dark-haired, olive-skinned women in line behind me seemed to be South Asian. But otherwise they belied the traditional stereotype of Muslim women. They were assertive and poised, rolling their carry-on bags behind them as they stepped up to the counter, throwing me friendly smiles as they passed. Their short hair danced on their shoulders without cover. After I got my room key, a buoyant teen guided me to the conference registration table. Her parents were conference organizers, and she was part of the Ismaili youth group volunteering over the weekend. My head was spinning seeing Muslim women like me and Muslim girls like the girl I used to be.

In my hotel room, I talked to my father to let him know we'd arrived safely. Before I knew it, I started getting angry. I hadn't even participated in the first session of the conference agenda, but I'd gotten an early sense of how much this community of Muslims valued women as vibrant, equal participants. I wasn't naive—I knew every community had its problems—but the openmindedness I'd seen in Ismaili Muslims was a far cry from what I'd experienced in my local Muslim community. "It's just ridiculous the way you treat women. We don't even get to speak," I yelled at my father. "We have to sit like we're deaf, dumb, and blind." "I know, honey," my father said. "I have fought with them so much. I am tired. They don't care."

As I talked with my father, Shibli played by my feet, fascinated by the white puffs of cloud that decorated a rug the bellhop had brought us to accompany a cozy crib. After I got off the phone, I danced with Shibli in our room, with lines of twinkling car lights from an Atlanta road shining behind us. High up in the sky in this high-rise building, I felt invincible and happy in a Muslim community in which I had a sense of belonging and worth as a woman.

The next morning I sat in the second row of the grand ballroom at the Atlanta Westin, meeting up with Amir when he waved me to the seat beside him. Upstairs in room 1202, babysitters were watching Shibli for the first time since the night eight months earlier when I threw rocks at the devil on the hajj and left him with fellow pilgrims. He was happily playing with his babysitters when I left.

I felt affirmed there in Atlanta among the Ismaili Muslims. The women didn't cover their hair. They sat anywhere they wanted to sit. The night before, a young woman, Salima, had worn a sleeveless black pantsuit without shame. Years before, a first cousin had scolded my parents for letting me wear my sleeveless Morgantown High track uniform, because a woman's biceps were not to be seen in public. According to the puritanical Muslims, a woman's upper arms are part of her *awrah*—the parts of the body that need to be shielded from public view. But there amid the Ismaili professionals, these rules didn't matter. Greater priorities defined this community, such as improving health care, education, and prenatal care.

In the ballroom I felt such anger toward the leaders of our Muslim community in Morgantown, including my father. Since the arrival of Arab immigrants with their strict cultural sense of segregation, we hadn't even been able to enjoy Muslim holidays and social gatherings as a family. There in Atlanta men and women sat shoulder to shoulder without judgment. A film rolled with footage of the Aga Khan's wife. A PhD, she wore a black sleeveless dress.

The morning's program started with the national anthem. I sang the words with my right hand over my heart. For most of my life I had not put my hand on my heart when I heard these words. I had always been made to feel embarrassed about being American when I went to India for summer trips. It was in the way my cousins made fun of my Urdu, my American accent, and even the fact that I ate with a fork instead of my hands, as is done in traditional culture. When I became a U.S. citizen in 1981, it was really only to fulfill the citizenship requirement for college scholarships. Then, in my late twenties, I found myself standing in the outfield bleachers above the Big Green Monster at Boston's Fenway Park, my hand over my heart, belting out the words to the national anthem. I was starting to realize that I was American.

Unlike in Boston, in Atlanta we stayed standing for a recitation from the Qur'an. "We have created you male and female," a woman said, her white *dupatta* falling off her head onto her shoulders without reproach. I had been taught that I couldn't read the Qur'an without covering my head for modesty. The Moroccan scholar Fatima Mernissi told me in Morocco that she didn't cover up every time. The Qur'an, she explained, was "a research book" for her.

With the formalities complete, it was time for the main speaker. She was Princess Zahra, a daughter of the Aga Khan and a 1994 graduate of Harvard with a degree in Third World development studies. She sat at the table with her brothers, Prince Rahim and Prince Hussain, working together for the Aga Khan Development Network, a group of institutions dedicated to improving living conditions and opportunities in the developing world through such ventures as micro-credit loans to Muslim women in Mozambique.

With her introduction, Princess Zahra strode to the podium with the determined gait of the athlete that she was, her calves bulging beneath the knee-length hem of her business suit, which was cut to fit her figure snugly. Her hair was pulled back into a tight bun, and her face glistened under the bright stage lights. With her regal air, Princess Zahra is the closest thing Islam has to a Princess Diana. She was even married in England to a British man who didn't convert to Islam. As she talked she exuded the kind of confident energy that I wished I could express in my Muslim community or even see in another Muslim woman. In her speech she laid out an aggressive and smart development policy for countries in the developing world.

The princess made an important point: education has a key role to play in emancipating Muslim society from traditions and ideologies that contradict fundamental Islamic beliefs regarding women, non-Muslims, and community. Toward that end, the Aga Khan Foundation had built Centers of Excellence in the far corners of the world, from Karachi, Pakistan, to the Central Asian states of Kyrgistan and Tajikistan and the African nations of Mozambique and Kenya. As she closed, I joined the applause, enthusiastically.

I left wondering how the global Muslim world, not just this pocket of it, could embody some of these principles of tolerance, women's rights, and plurality, all rooted in Islam. I went to meet, for the first time, the Muslim scholar who had helped me find so many answers to my questions about my faith. He was Alan Godlas, the professor of Islamic studies at the University of Georgia. Its campus is in Athens, just outside of Atlanta. Not only was I meeting him for the first time, but so was Shibli. They took to each other immediately. "This," I told Shibli, "is your godfather who will teach you Arabic." I wanted to help my son to understand his religion fully and represent it, as much as he wanted, as positively as he could in the world. Shibli smiled warmly at Dr. Godlas, and their friendship was sealed.

As we played in the nursery, chasing Shibli here and there, I asked Dr. Godlas the question on my mind. Referring to the questions I raised in my first book about the Muslim world's relationship to terrorism, intolerance, and the criminalization of women's bodies, he said, "Your book is the kicking of the fetus in the belly of Islam." I looked at him curiously. He continued. "One can look at you and Islam together as the fetus kicking and trying to come out. Islam wants to be born into the postmodern world." He confided that he had felt his own anguish as he read my book. "There are still forces trying to hold this fetus inside. It hasn't grown enough, and the mother who is in the world is the collective consciousness of Muslims. The fetus is the growing soul." I knew this baby had to be born.

A MOTHER'S COMMITMENT TO ISLAM

MORGANTOWN—As Shibli grew, I didn't really know if I believed enough in my faith to initiate Shibli into it. After all, some Muslims invoked Islam to label me a criminal. For that reason, I couldn't throw a naming ceremony, called the *aqeeqa*, which Muslim parents usually host for the Muslim community days after a baby's birth. But my heart changed through the summer as I toured for my book and read the words of support from the Muslim women and men who wrote to me after I wrote in defense of Amina Lawal. They made me believe that maybe I could find a Muslim *sangham*, a sort of spiritual community in Buddhism. Maybe I could commit to raising my son a Muslim. With these hopes, I scheduled my son's aqeeqa with his first birthday party.

The crispness in the autumn air reminded me of how far I had come in one year with the help of my sangham. They had given me something that the Buddhist teacher Tara Brach told me Buddha called the "wings of awareness." I had called Tara to understand the spiritual power of my friends who took care of me when I returned dejected from Pakistan, via Paris. "They recognized your vulnerability and held it with compassion. They saw your goodness, and they acted as a mirror for it," Tara told me. "The common denominator is unconditional acceptance."

My sangham included my friend Pam Norick, who introduced me to Tara's latest book, *Radical Acceptance*, which contains the kind of philosophy that she thought could help me. I had met Pam in the summer of 1986 when we both started as graduate students at the American University in Washington. Pam went to Congress as a foreign policy and defense adviser and worked there for nearly a decade before starting her own consulting business specializing in global health and development. As someone who grew up in a Catholic family and then charted her own path for a life as a spiritual activist, she has been a gentle muse for my growth as a woman. Pam comes from a family that up until her generation always "gave" one girl to the Catholic Church to serve as a nun. Seeking a different path, college, and serving for two years as a teacher in Zaire in central Africa, Pam always cultivated a strong spiritual life. She learned to meditate when she was nineteen, a practice that has grown even more central to her life in the past twenty years. She takes silent retreat several times a year. "I believe that silence is a great teacher," she told me, "and instructs us to use our whole minds—our intuition and our reason—so that we can live our lives more consciously and with the welfare of others always in mind."

Another friend was Rachel Kessler. Effervescent and sincere, she delivered to my home, years earlier, a pile of books on depression so that I could understand the gloom I was experiencing after my divorce. We met at the Washington office of the *Wall Street Journal* when Danny and I worked there. Born in New Rochelle, New York, Rachel had a mother who was raised Polish Catholic and converted to Judaism when she married Rachel's father. After a divorce when Rachel was six, her mother slowly reembraced her Catholic upbringing and brought Rachel and her brother to midnight mass on Christmas. In high school Rachel struggled with the question of whether she was Jewish or Catholic. Her father made sure she understood her roots, and she found particular inspiration in her paternal grandmother, Minuetta, an independent woman who as a composer worked in a profession made up mostly of men at the time. Minuetta broke from tradition and divorced her husband, a renowned scientist, in 1949, when it was particularly unheard of for a woman to be a single mother. "Why do I support you?" Rachel said one day when I posed the question to her. "Perhaps because I was so close to my grandmother, a pioneer in her own way, as her mother was a strong pioneering woman as well. Perhaps it's because I come from a family that believes in living life fearlessly, in enjoying and indulging and working hard. Perhaps it's just because that's what friends do." Her words touched me.

My guests at Shibli's party were from the sangham that had embraced me both locally and from afar after I returned home from Karachi. In Morgantown, Koula Hartnett, a Greek American, and her husband, Richard, walked their dog, Prince, every night by my parents' house, dispensing kind words and support when they saw me outside. My college friend Ellyce Johnson drove in from Pittsburgh for the party. A devout churchgoer, she had a buoyant spirit that kept my spirits up. From Washington, D.C., my dear friend Lynn Hoverman came to Shibli's party. She was a thirtysomething, blue-eyed American woman of German Lutheran ancestry on one side and Midwest Methodist lineage on the other, born and raised in southern California in the town of Lancaster, a high-desert region of the Mojave Desert where tumbleweeds blew, high school pride was strong, and Little League and rodeo reigned. She was baptized in the Methodist Church, and a girlhood of Sunday school and confirmation classes followed. Her Methodist church had only male ministers, but both men and women taught Sunday school classes, and she attended other Methodist churches where female ministers led the services. Boys and girls equally shared the honor of lighting the altar candles, and the church had openly gay members. Unlike at the churches of some stricter Christian faiths that she attended with childhood friends, Methodism was never touted as being better than other forms of worship. After 9/11, she said, "I began to understand that only moderate voices speaking up and speaking up loudly could counter the extremists. Any real change has to come from within Islam itself."

For this occasion I wanted to commemorate Shibli's name, "my lion cub." I found just the right symbol down Route 7 from Morgantown at an unusual place called Hovatter's Zoo. A tiger there had just had a new litter of babies. The owner would send a tiger cub for Shibli's birthday. In a parade befitting a lion cub, we marched together, the unlikely mix of people who made up my sangham. And in the jungle that was our backyard, Shibli and his friends petted this tiger cub that had come to help my lion cub celebrate his birthday.

Up the hill, we gathered on our back deck for the symbolic moment that had brought us together: a hair shaving that symbolizes a baby's initiation into Islam. To continue would be, for me, a conscious decision to raise Shibli a Muslim. I paused to reflect. What I hoped for Shibli was that he would be as stellar as all those in my sangham—Christians, Jews, Buddhists, Pagans, and Hindus—who strove to live honestly and well. I would teach him the rituals of my faith, as I did on the hajj, and guide him to understand that, while our paths with others might be different, they were parallel paths toward goodness.

My friend Nancy Snow, a scholar in propaganda, had come to Morgantown for Shibli's birthday, and I found an applicable lesson in her vocation. She had studied with me when I was at American University, and she had made it her life mission to dissect how political and religious leaders so often demonize others to promote their own agendas of power and persuasion. Growing up, she had gone to a Southern Baptist church and heard weekend after weekend that only those who took Jesus as their personal savior would go to heaven. In many conversations, she told me, as part of the peace-loving moderate majority, it was the responsibility of each one of us to avoid getting manipulated by the propaganda spewed from the pulpit. She had gone to the Christian fundamentalist Bob Jones University. "Bob Jones taught me at the time to see the world in absolutist parameters—good versus evil, Protestant versus Jew, and Muslims weren't even discussed. As a result, instead of looking at the world through rose-colored glasses, I looked at the world through a continual question of 'Are you with me, or not?' The *not* was anyone who wasn't Christian on my terms; you were sadly classified as headed toward the going to hell exit," she told me.

When Nancy later studied propaganda, she saw how closely it was linked to messianic visions for absolutist adherence to a particular belief system. In 1622 Pope Gregory XV started the Congregatio de Propaganda Fide—the Congregation for the Propagation of the Faith. Sure enough, *propagation* was the same word the Saudi Wahhabi literature used. "My Bob Jones days trained me in the art of conversion, but now I'm not trying so much to convert as to get us all collectively to think harder and longer about the enemy images we engage in, how words and visuals can both harm and heal, and our need to teach love and mercy as much if not more than fire and brimstone damnation." The puritans of all religions try to divide and conquer the world with arrogant interpretations of faith that give them a monopoly on the path to salvation. Looking out at my friends, I knew the puritans were wrong.

With that, my father slowly shaved Shibli's hair, just as the prophet had done with his daughter Fatima's son, Hussain, seven days after his grandson's birth. To the surprise of everyone, Shibli sat quietly on his rocking chair. I had done it. I had made a serious commitment with this simple act. I had no idea, at that moment, what it would mean in my life. I was just happy that Shibli didn't get a scratch from the ceremony.

As our last guests left I reflected with Ellyce and Lynn on the events of the past year. Shibli's conception hadn't followed the script that I had planned for myself, but I accepted the responsibilities that came from my life choices. I embraced my son and dedicated myself to raising him with virtue, respect, and honor. It had taken me a year, the hajj, and my tour of liberation behind me to decide to give my son his aqeeqa and raise him as a Muslim. Now that I had made that choice, I knew I had to find a way to coexist peacefully with the people of my Muslim community.

PART SIX

ASSERTING THE LESSONS OF THE PILGRIMAGE

October 2003 to May 2004

REALITY CHECK AT MY LOCAL MOSQUE

Oh ye who believe!
Stand out firmly
For justice, as witnesses
To God, even if it may be against
Yourselves, or your parents
Or your kin.

"Al-Nisa" (The Women),
Qur'an 4:135

MORGANTOWN—For most of my life I quietly bypassed traditions instead of directly challenging them. When I was married in Islamabad, Pakistan, an uncle told me that my grandmother couldn't be a witness. Two women equaled one man as a witness, according to Pakistan's sharia. Neither relenting nor opposing the edict, I had my grandmother *and* my aunt serve as one witness.

I had always distanced my life a bit from the Muslim community. After 9/11 and Danny's murder, I recognized that the stakes were huge for how Muslims expressed themselves in the world. Muslims like me sat silently while militants wrenched the religion from us and declared they were the protectors of the faith. My immersion into darkness and my experience in the light of the hajj transformed me. It made me recognize that I have a role in standing up to the extremists in my religion who try to intimidate us into respecting and following them. The purpose of religion is to inspire in us the best of human behavior.

Having chosen to raise Shibli as a Muslim, I wanted to find kinship in my local Muslim community. The opportunity came two weeks after Shibli's first birthday. The board of trustees at the Islamic Center of

Morgantown was opening a new mosque that had cost approximately $500,000 to build. It was a new three-story brick building on a street off the campus of West Virginia University, not far from a McDonald's and Hartsell's Exxon. It had a large community room, a kitchen, a small library, and a small office on the first floor, a massive prayer hall on the second floor, and a small balcony on the third floor. The afternoon before the opening a local West Virginia woman, a convert, called me to ask for my help in making sure that leadership set the right precedent for women's rights at the mosque. She had stood in the parking lot across the street from the old mosque when the men claimed there was no room inside for women. She and another convert were turned away from the door when they went to the old mosque to join in breaking the fast during the holy month of Ramadan. She had organized private swimming for Muslim sisters on Sunday nights at an indoor pool, even taping newspapers to the windows, but puritanical gossipmongers sniped at her efforts. And she had organized a family day at Valley's Worlds of Fun, an entertainment center with laser tag and other games, but an extremist protested an activity called "Rock and Roll Cage," thinking it would promote improper Western dancing. In fact, it was an innocent four-person ride.

These conflicts reflected the imposition of cultural hang-ups on the community, but I realized that there was a much more serious threat to the community than the exclusion of women and canceled activities. Years earlier an Iranian American professor of international communications at American University, Hamid Mowlana, taught me that the mosque plays an important role in defining and disseminating Islamic ideology in the world. He identified the mosque as a place that serves not only as a site for daily prayers but also as a communications model for spreading news and opinion and as a forum for political decisionmaking. My friend Danny's kidnappers had used a mosque in Karachi as a drop-off point for the photos the world saw of Danny in shackles. I felt it was imperative to make mosques—not to mention all places of worship—safe refuges for men, women, and children, not safe houses for hatred, division, and puritanical ideology.

I agreed to meet the West Virginia convert at the McDonald's by the mosque, and I drafted a seven-point "Manifesto for Equal Participation by Women" that included equity in access, accommodations, facilities, and services. The women's restroom, for example, included no footbaths, while the men's restroom had spacious facilities for *wudu*, the ritual cleansing mandatory before prayer. My father joined us and took the manifesto to a meeting of the board that night. He reported to me that they didn't respond, but they listened to the points. With hope, I walked with great enthusiasm to the freshly painted green doors of the new mosque. It was the eve of Ramadan, the holiest of months for Muslims and a time when we abstain from food, drink, and sex from sunup to sundown on the path to liberate us from our attachments to worldly desires. I had enjoyed this month since my childhood days as a sort of spiritual boot camp. I wore the same flowing white hijab I had worn in Mecca, Medina, and Jerusalem.

"Sister, take the back entrance!" the board president, an Egyptian American man with a PhD, yelled at me. I was stunned, not only by his message but by his tone. It had none of the warmth I'd received from not just women but also men in Mecca and Jerusalem. He didn't even give me the typical Muslim greeting, "As-salaam alaykum." I was taken back to Mina, where I had learned that small acts of kindness can mean so much.

He expected me to take a wooden walkway along the right side of the building to a back door. It opened into a back stairwell that led to an isolated balcony considered the "sisters' section." I was so stunned that I continued through the front door, but I didn't dare go up a set of stairs to the right, inside the front door. They led into the main sanctuary. Somehow I knew it was off-limits to me as a woman. I walked through the community room and slipped into the back stairwell and climbed into the balcony. Resentful, I sat down on the carpet, cross-legged, Shibli on my lap. I felt humiliated and marginal. I stared at a half wall. I couldn't see into the main hall unless I looked over the edge, and I—a woman who had fearlessly crisscrossed the globe, meeting with heads of state, the Taliban, and chief executives—didn't have the nerve to even look over the wall. I heard the disembodied voices of prayers and lectures from the main hall downstairs, but with the distance, the inadequate sound system, and the chatter of women who were socializing because they were disconnected from the mosque's main activities below, I could barely make out what they were saying. I didn't have the nerve to speak up or protest. If I had something to say, I was supposed to write my thoughts on a piece of paper, pluck some child from play, and ask the child to pass my note to the men in the main hall. I felt like a second-class citizen.

Worse yet, I was confused. I had never been treated so rudely at the Sacred Mosque in Mecca or in the Holy Sanctuary in Jerusalem. I had walked through those gates freely. I had navigated the halls without restraint. In Mecca I had emerged into the courtyard that housed the Ka'bah without interference. Even though I opposed most of Saudi Arabia's policies toward women, the government made the hajj experience

more equitable than I could have ever imagined. The Saudis hadn't
forced families to separate to observe the usual strict segregation that the
state imposed on men and women in the public sphere. And yet it was
unacceptable for me to walk through the front doors of my own mosque in
Morgantown, West Virginia. I hadn't attached much significance to the
moment when I had stood before the Ka'bah with my family, unhindered
by gender segregation, but in a lesson I was slowly learning, I realized
there really were no moments from the hajj that were without meaning.

I went to my parents' home angry. I yelled at my father in the kitchen.
"Who decided women are so worthless?" He shook his head, sharing my
frustration but feeling helpless. We were lucky that we hadn't been con-
signed to a totally closed-off room. The board originally had accepted a design in which the wall overlooking the main hall would have been
floor-to-ceiling. My father protested that it would cut women off too much.
The board wouldn't abandon the design until the fire marshal told the
board that it would have to pay thousands of dollars for a sprinkler system
if it kept the balcony barricaded. It was an uphill battle to assert women's
rights in the Morgantown mosque. "They're male chauvinists," my father
said. I tried to accept the status quo through the first days of Ramadan,
praying silently upstairs, listening to sermons addressed only to "brothers."

On Friday the discrimination became really obvious. The community
was gathering to break the fast at a sunset meal called *iftar*. A new female
medical resident at the university arrived at the new mosque for the iftar
dinner. Enthusiastic about being a part of her new Muslim community,
she walked up to the front door with a covered dish. A man wouldn't
even look her in the eye as he issued his order. "Women—over there!" he
yelled, pointing to the old mosque. "This is the opening of the new
mosque. I want to enter," she protested. But how could she penetrate the
phalanx of engrained cultural misogyny and discrimination that made the
new mosque a men's club? She slunk over to the cramped old mosque,
where a posse of kids ran wild.

The women got a dingy room in the old mosque. We sat on white trash
bags on the floor next to ratty filing cabinets that seemed ready to topple
over. It was so cramped that my plate of food tumbled to the floor when I
stood up as a boy ran by me in play. I went outside and stared at the
brightly lit new mosque across the street. I could see the chandeliers inside sparkling in the main hall. The men streamed in and out of the
mosque, happily enjoying their dinner inside the spacious community
room. We were not included in their definition of community.

Fed up, my mother dispatched a boy to get my father. On the ride home my mother expressed her frustration. "If I was a man, I couldn't even eat my dinner if I was getting the best while the women and children were getting the worst. I would take the worst and give them the best." I thought of my nephew, Samir, in Medina refusing to see the prophet's original mosque when his sister couldn't. This was the kind of enlightened thinking we needed to awaken in our men. Not a single one of them should have accepted their entitlement to the spanking new mosque knowing their wives, daughters, sisters, and children were being denied access to it.

Despite my outrage, I felt I would be like an interloper if I protested. But my awareness of our subjugation interrupted my prayer each time I touched my forehead to the carpet. I lay in bed each night despising the men who had ordered me to use the mosque's rear entrance and questioning the value my religion gave women.

I called the scholar Alan Godlas at the University of Georgia for guidance. He empathized immediately with me. He said it wasn't Islamic to treat women as the men in my mosque were doing. But he also knew that the wound I was feeling ran deep. "Your anger reveals a deeper pain," he said. Over the next days I wondered about what he had said, and I realized it was true. I had witnessed the marginalization of women in many parts of Muslim society. But my parents had taught me that I wasn't meant to be marginal. Nor did I believe that Islam expected that of me. I began researching that question. I wanted to know the truth.

HISTORY RECLAIMED

MORGANTOWN—I found overwhelming scholarly evidence that mosques that bar women from the main prayer space aren't Islamic. They more aptly reflect the age of ignorance, or Jahiliya, in pre-Islamic Arabia.

Through the power of our information age, I discovered that I wasn't alone in my rebellion. Working by day in my childhood home in West Virginia, I found Asma Afsaruddin, a professor of Arabic and Islamic studies at the University of Notre Dame. "Women's present marginalization in the mosque is a betrayal of what Islam had promised women and [what] was realized in the early centuries," she e-mailed me, adding:

There is solid evidence in the early Islamic literature (in collections of *hadith*, for example) that women were not relegated to the back of the mosque to pray during the prophet's time or cordoned off from the main congregation. The customary practice of doing one or the other in mosques today represents significant shifts in attitudes towards women's presence in the public sphere when compared to the first century of Islam and reflects the influence of local culture and customs on religious praxis. The sources at our disposal clearly show women of the first generation of Muslims in particular participating avidly in the communal, religious, and political activities of their time. Women were transmitters of religious knowledge, providers of social services, participants in battles and in political affairs in the early period. By the third century of Islam, many of these rights slowly began to be whittled away as earlier Near Eastern—primarily Sassanian and Byzantine—notions of female propriety and seclusion began to take hold. However, by no means, did female activity in these spheres cease.

The marginalization of women seemed, if anything, to be worsening. The Islamic Society of North America, the national Muslim organization whose hajj tour group I had joined, conducted a survey in 2000 in which it concluded that "the practice of having women pray behind a curtain or in another room is becoming more widespread" in the United States. In 2000 women at 66 percent of the U.S. mosques surveyed prayed behind a curtain or partition or in another room, compared with 52 percent in 1994, according to the survey of leaders of 416 mosques nationwide.

The survey revealed a dearth of women in leadership positions in many mosques. Over the past five years women had served on the boards of about half of the mosques. Shocking to me, they weren't even *allowed* to serve on the boards of 31 percent—or one in three—of the mosques surveyed. If parent-teacher associations anywhere had similar bans on Muslims, it would be appropriately unacceptable. It was unconscionable to me that national Muslim organizations had known for two years about the gender discrimination uncovered by the survey and had aggressively not done something about it.

I didn't know where the modern-day scholars fell. I went back to the Islamic Society of North America and looked up its vice president, Ingrid Mattson, the Islamic scholar at the Hartford Seminary who had spoken at the Islamic Society's annual convention in Chicago. I thought she would be as good a person as any to test for official Muslim world opinion. To my surprise, I didn't have to explain anything to her. All too often the mosque in America "is a men's club where women and children aren't welcome," she told me. Afterward, I double-checked my notes indeed called mosques "men's clubs."

ISNA is affiliated with an outfit called the Fiqh Council America, which issues its rulings on questions of *fiqh*, or jurisprudence. I was stunned to discover among its rulings a fatwa supporting women's rights in mosques. "It is not required to place a curtain or division in the mosque between men and women. There is no verse in the *Qur'an* or *Hadith* of the Prophet—peace be upon him—that tells us that we must do so," wrote Muzammil H. Siddiqi, a Fiqh Council member and past president of the Islamic Society of North America. He concluded that it is perfectly Islamic to hold meetings in the mosque that include both men and women, whether for prayers or for any other Islamic purpose, without separating them with a curtain, partition, or wall. Siddiqi further noted: "In America we are living in a very mixed society. Our brothers and sisters are all going out for work, shopping, study or just for outings. Our mosques should be the places where we should learn and teach our children the etiquette of living in a mixed society with Islamic manners. If we make artificial barriers between men and women inside our places of worship, where are we going to learn Islamic manners of being together as believing men and women?"

I was breathing easier and easier as I started to realize that the oppressive spirit I so often felt in my Muslim community was denying essential rights that Islam had granted women in the seventh century. I learned too that, besides Dr. Mattson, there was an entire new generation of Islamic scholars who were ready to dismantle the traditional barriers to the full realization of women erected by men. One academic led me to another in this loosely knit network of scholars who were like a secret society of men and women dedicated to resurrecting the early principles of Islam, among them the power of the feminine. A monk named Martin Luther had posted reforms of the Catholic Church on the doors of a church in Wittenberg, Germany. To me, Muslim women scholars were the Martina Luthers of the Muslim world, and men like Dr. Omid Safi and Michael Wolfe were humble Martin Luthers publishing important books in our modern-day effort to return Islam to its essential teachings. They were not trying to create a schism but rather to take back Islam, the title of a book edited by Michael. In Ithaca, New York, Islamic scholar Asma Barlas wrote *Believing Women in Islam: Unreading Patriarchal Interpretations of the Qur'an*. A year earlier I would never have read anything with the word *patriarchal* in its title. Now I took it every night with me to the mosque. Ramadan is the month when it's said God started revealing the Qur'an to the prophet Muhammad. We were supposed to spend the month reading the Qur'an. It was the right month for me to understand how men had interpreted the Qur'an to deny women's rights. It was time that we took back our religion.

The systematic elimination of women's rights was starting to make sense to me. I was relieved. It was about power and control. It wasn't about Islam.

From Richmond, Virginia, a scholar by the name of Amina Wadud told me that the Islam concept of *tawhid*, or the oneness of Allah and all beings, emphasizes the equality of men and women. An Islamic studies professor at Virginia Commonwealth University, she coined the term "tawhidic paradigm" to assert women's equal rights at the mosque.

The Qur'an ("Al-Ghafir" [The Forgiver], 40:40) evokes the virtue of righteousness equally in men and women.

> Whoever does an atom's weight of good, whether male or female, and is a believer, all such enter into Paradise.

In "Al-Imran" (The House of Imran, Qur'an 3:195), God says that men and women are equally rewarded for their acts of goodness on this earth:

> I will deny no man or woman among you the reward of their labors. You are the offspring of one another.

Over and over again the Qur'an teaches us that women are equally charged with carrying their weight. It was a lesson that I learned on a reporting assignment in the summer of 1984 outside Washington, D.C. I was a nineteen-year-old summer intern at a sweatshop called States News Service, and I was wearing night goggles while going through field exercises run by the Army National Guard in light infantry warfare, the kind of military strategy that the United States has employed in Afghanistan and Iraq. "Everyone carries their own ruck," the army sergeant barked, referring to our backpacks. I took his point to heart and saw that the Qur'an imparts this idea to us as citizens of the world.

> Each person shall reap the fruits of his/her own deeds: no soul shall bear another's burden.
>
> "Al-Anaam" (The Cattle),
> Qur'an 6:164

The true believers, both men and women, are friends to each other. They enjoin what is just and forbid what is evil; they attend to their prayers and pay the alms and obey God and His apostle. On these God will have mercy. He is Mighty and Wise.

> "Al-Araf" (The Heights),
> Qur'an 7:71

Dr. Wadud claimed that the rights Mecca gives women, including the right of women to pray together with men at mosques, are affirmed in something she refers to as "Meccan *salat,*" or prayer without formal gender segregation, as we had experienced in Mecca. I related to Dr. Wadud on so many levels. She was a single mother and a Muslim feminist. After we got off the phone, she sent me an essay extolling the virtues of Hajar.

My eyes were opened to see Muslim women excelling in the world under the umbrella of Islam. I learned about an Indian Kashmiri Muslim woman in New York with an unorthodox first name, Daisy Khan. She had met her husband, a bold Muslim leader by the name of Imam Feisal Abdul Rauf, in the mosque he helped lead, Masjid al-Farah. I connected with her immediately when I called her. "Your marriage is what happens when genders mix in the mosque," I half-joked when she told me her story. What had happened was good. Together she and her husband were dedicated to expanding the traditional definition of our Muslim community.

Daisy Khan had responded to the tragedy of September 11, 2001, by sponsoring interfaith art exhibitions through an American Muslim organization she had founded with her husband, the ASMA Society. She told me that women had a designated space in their mosque. Late-arriving men often ended up praying behind the women already settled in their space. Her husband even had women do the sacred duty of the call to prayer. And Daisy led mixed-gender study circles. She told me, "The mosque is a place of learning. . . . If men prevent women from learning, how will they answer to God?" She made such sense. She also was a tough cookie. When a woman at a mosque tugged at her sleeves while she was in the midst of prayer, to admonish her to pull her sleeves to her wrists, Daisy turned to the woman afterward and said, "How are you going to answer on your judgment day when I testify to God that you prevented me from doing my prayer?" The woman pleaded, "Please forgive me! Please forgive me!" I made a mental note to keep that tactic in my bag of defenses against theological intimidation.

Much of the discrimination of puritanical Muslims against women is practiced in the name of "protecting" women. If women and men are allowed to mix, the argument goes, the mosque will become a sexually charged place, dangerous for women and distracting to men. In our mosque only the men were allowed to use a microphone to address the faithful. The first Friday of Ramadan, the West Virginia convert who had organized Muslim sisters' swimming excitedly announced that a student, Dana, would recite her *shahada* for the first time. For all Muslims, the first and most important pillar of Islam is the testimony of faith, or shahada: "There is no true God but Allah, and Muhammad is the messenger of Allah." To be a Muslim, you have to say and believe these words. They are supposed to be the words we utter with our dying breath. To hear these words uttered by a new convert is considered blessed.

The West Virginia convert asked the mosque manager to bring the microphone to the women's balcony so that everyone upstairs and downstairs could hear the student's shahada. He refused, telling her: "A woman's voice is not to be heard in the mosque." What he meant was that a woman's voice—even raised in prayer—is an instrument of sexual provocation to men. Some puritanical Muslims consider a woman's voice *awrah*, or a part of a woman forbidden to be displayed publicly. Many women accept these rulings, and their apathy makes them the status quo. The women in our mosque accepted the ruling. The men downstairs talked noisily through the conversion ceremony, oblivious to the profound act upstairs of Dana converting to Islam. Dana left, disheartened. "This isn't the Islam I was promised," she said, slipping out by the darkened rear staircase for women.

One of the issues working against American Muslim women—an issue not much discussed outside the Muslim community—is the de facto takeover of many U.S. mosques by puritanical and traditional Muslims, many from the Arab world. I had gotten a sense of this ideological shift from my father's comments about his frustrations at the mosque. It was the convert who had urged the writing of the women's manifesto who identified the takeovers as ideological. "They're Wahhabi and Salafi," she told me quietly one night.

Wahhabi, I understood. It's the ideology that defines Saudi Arabia and restricts the lives of Saudi women, for instance, by forbidding them to drive. But *Salafi*? I turned to the modern tool for empowerment: Google. What I learned helped me connect the dots. Salafism is an ideology related to Wahhabism. Its proponents fancy themselves as "pioneers," borrowing on the word *salaf*, which means the early generations of Muslims, and, like Wahhabism, it follows a very narrow interpretation of the Qur'an and the Sunnah—the sayings, traditions, and examples of the prophet. Fundamental to its interpretations is male control over women.

Now I got it. Since my childhood days, immigrants who followed this ideology, most of them from Egypt and Saudi Arabia, had overrun my town, and they had managed to transform my local Muslim community through strict rules of segregation and disempowerment of women. Even though the man who had most liberated my mind to think freely, Dr. Mahmood Taher, was also from Egypt, what I learned was that Muslims are in an ideological war: just as in the United States, there are many competing strains of puritanical and moderate thinking in Egypt and Saudi Arabia and elsewhere. The immigrants at my mosque had stacked our library with Wahhabi books printed by the government of Saudi Arabia. To my shock, the local Muslim Students' Association distributed a book, *Women in the Shade of Islam*, by a Saudi sheikh, Abdul Rahman Al-Sheha, that included a section called "Women Beating." He repeated a disputed translation of the Qur'an that the puritanical says allows "beating of women . . . in restricted and very limited occasions [such as] when a wife disobeys her husband's instructions for no visible and acceptable valid reason." The sheikh noted that "submissive, or subdued women" are the kind who "enjoy being beaten."

There were virtually no students from Saudi Arabia and Egypt in Morgantown in 1993. More precisely, there were three. By the fall semester of 2003, however, there were fifty-five, mostly male and conservative, and their wives regularly glided through the local Wal-Mart wearing black *abaya*, the gowns I'd seen in Saudi Arabia that cloaked their faces as well as their hair in black. I was starting to understand what had happened.

Sadly, the students' presence emboldened (or in some places cowed) American mosque leaders, many of whom tried to rationalize discrimination against women through a hadith: "Do not prevent your women from [going to] the mosques, though their houses are best for them." Scholars consider this hadith an allowance, not a restriction. The prophet made the statement after women, busy with household chores, complained when he said Muslims get twenty-seven times more blessings when praying at the mosque.

As I continued my Internet sleuthing, I found a document that didn't make sense. It was a fatwa by the Saudi government stating that partitions and separate rooms aren't required in mosques. I called the Saudi embassy in Washington to authenticate the decree. I found the press spokesman, Nail al-Jubeir, brother to the aide to Prince Bandar, Saudi

Arabia's ambassador to the United States. Talking fast and feverishly, he confirmed that the ruling was true. Unprompted, he brought up the example of Muslim women and men praying freely at the Sacred Mosque in Mecca. I was stunned. The government of Saudi Arabia understood my point? It agreed with me? Maybe the Saudi government was practicing post-9/11 spin control to distinguish itself from militant Muslims like Osama bin Laden and the September 11 hijackers. Whatever the Saudis' intention, I thought to myself, *I'll take it.*

I told Nail al-Jubeir about life in our mosque and explained that many of the members were Saudi students arguing for the strictest separation for women, essentially marginalizing us. "You know what I tell these students?" he said. "They had a choice to come to study in America. They could have studied in Pakistan or one of the Gulf countries. But they chose America. If they don't like it here, I tell them, 'There are three flights a day back to Saudi Arabia.'"

I couldn't believe my ears.

On the tenth night of Ramadan I went to our student union, the Mountainlair, on the campus of West Virginia University. I had zipped through it many times a day as an undergraduate, shuttling between classes and my interviews as a hard-charging reporter for the student newspaper, the *Daily Athenaeum.* As in the rest of Morgantown, this was a place where I had fully realized my potential. The West Virginia woman convert had booked the theater to show the documentary *Muhammad: The Legacy of a Prophet,* coproduced by the Muslim writer Michael Wolfe. The theater was mostly empty: men from the mosque were packing a room down the hall to discuss their civil liberties under the Patriot Act. Muslim men had been detained and deported under new powers the federal government had given law enforcement after September 11. I deplored the abuse of civil liberties, and sitting in the theater I saw how our Muslim world had also eroded. Watching the documentary, my heart wept for how far Muslim society had strayed from the ideals of Islam espoused by Muhammad, and I had an epiphany: Muslim women, with Muslim men supporting them, should obey the Qur'anic injunction to fight *zulm,* or oppression, even if it came from within our own community.

In the courtyard of the Sacred Mosque in Mecca, I prayed just about fifty feet from the Ka'bah with a swirl of women *and* men beside me, in front of me, and behind me. There were no partitions, barriers, or curtains. All over the world Islamic culture keeps men and women from praying next to each other in mosques. But as I'd seen, most American mosques, like mine, have gone well beyond that simple prohibition by importing a system of separate accommodations that provides women with wholly unequal services for prayer and education. And yet, excluding women ignores the rights that the prophet Muhammad gave them in the seventh century and represents a *bida*—an innovation that emerged after the prophet died. I had been wrestling with these injustices for some time when I finally decided to take a stand as the credits rolled on the life of the prophet.

I had no intention of praying right next to the men even though I felt I had a right to at least a parallel section beside a section for men. To me, it was a compromise to accept praying behind the men. I understood I was accepting a second-class status from third class. I just wanted a place in the main prayer space. That night I sat at my computer and shared my epiphany with Michael Wolfe. "As I listened to the life of the prophet Muhammad, I wanted to weep thinking about the sorry state of the Muslim community today. We have wandered so far from the principles of fairness, compassion, and simple consideration that are supposed to be at the heart of Islam. We are not going to accept the discrimination any longer. It is ultimately the principles in Islam of equity and a woman's rights that guide us. My family is going to act together. We will pray beside each other as we did earlier this year in Mecca and Jerusalem in two of the holiest mosques in Islam. Our mystical night journey will continue to Morgantown."

OUR ASCENSION

Few will have the greatness to bend history itself; but each of us can work to change a small portion of events, and in the total of all those acts will be written the history of this generation.

Robert F. Kennedy (1925–68)

Make your land a sacred area
For you are in the sacred area.
Make your age a sacred time
For you are in the sacred time.
Make this earth into a sacred mosque
For you are in the Masjid al-Haram.
For "the earth is God's mosque"

And you see that:
It is not.

Ali Shariati (1933–77)

MORGANTOWN—In the predawn darkness of the eleventh day of Ramadan, my family and I continued the mystical journey we had begun when we traveled from Mecca to Jerusalem.

The prophet Muhammad had climbed onto a winged creature called the *baraq*. We climbed into my family's Chrysler minivan, my father behind the wheel. We set off on the wings of our commitment to the essence of Islam: truth, justice, and equality. We were the same ordinary pilgrims who had set off for Mecca almost eleven months earlier: my father, my mother, Samir, Safiyyah, and Shibli. We had prayed together as a family in the Sacred Mosque in Mecca and the Holy Sanctuary in Jerusalem; now I wanted to realize that experience at home. That morning, as the mist sat in the West Virginia hills like clouds descended to earth, we turned down a winding street called Van Voorhis Road and passed a branch of the local bank, BB&T, and a convenience store called Dairy Mart. My father nosed our baraq toward the new mosque, the road mostly desolate. Samir remembered the call to God we had recited as we approached the Ka'bah. "Labayk. Allahumma labayk. Labayk. La shareeka laka."

"What does it mean in English?" I asked the children. As we dipped down a hill I answered my own question: "Here I come. At Your service, oh Lord. Here I come. No partner do you have."

With my stomach churning as I worked to overcome my fears, my mother, my niece, and I walked through the moonlit front doors of our mosque with my father, my nephew, and my infant son. Resolute, we turned to the right and climbed the front staircase for the first time ever. It was only two paces from the front door, but we had never dared set foot there before. We quietly ascended and emerged into the greatest hall I felt I had ever seen. All the rules that forbade me to enter that space made it seem like a place larger than life. It was a space equal to a hotel ballroom, and entering it seemed as great an achievement as the prophet Muhammad's mystical ascension to the heavens. I had had to liberate myself from fear, control, and tradition to stand in this vast space. There were only about half a dozen men sitting in the front of the hall, a good thirty feet in front of us. We sat in front of the pillars lining the back of the great hall. Behind us was a darkened space, the women's balcony above it. "It's no big deal," Samir whispered to me, relieved.

Samir thought to himself, *This is it. There is no going back.* He was frightened—a little less than when he trespassed at the KFC in Mecca with his unescorted grandmother—but he felt he was taking the right stand. He had grown up on the tales of men and women in history fighting for their rights. He admired Martin Luther King ("He did things not only for his own ego but everybody"), Rosa Parks ("She thought everybody should have the same equal rights"), and Muhammad Ali ("The coolest Muslim I've ever read about"). He recalled how Muhammad Ali broke racial barriers to become heavyweight champion of the world. Samir had just finished writing a book report on a biography of "the Greatest." When Muhammad Ali was eleven, he'd discovered, members of the Ku Klux Klan stole his bike. That motivated him to learn how to box so that he would learn to overcome his fears. I told Samir I had met Muhammad Ali at a memorial for my friend Danny. Ali had responded to a request by Danny's father, Judea Pearl, to appeal to Danny's kidnappers not to kill him, and he came to the memorial to offer his condolences. He was feeble when I met him, but a strong symbol of a man fulfilling Islam's principles of tolerance. Samir knew what it meant to make a commitment. "I took a vow," Samir said, "to help the freedom of women."

To him, the logic was obvious. "It isn't fair for women to be alone, sitting alone, with the men clustered together somewhere else. If we're all in the same religion, why should we be separated?" As he looked over at my mother, his sister, and me, Samir felt good that the family could pray together. He saw no reason to leave his grandmother, sister, and me behind just so he could enjoy the vast main sanctuary.

Next to him, Safiyyah was the new pioneer Muslimah, the feminine form of Mussalman, or "Muslim." At school and at home, she got messages encouraging her sense of self-worth. When I talked with her about the arrangement at the mosque for women, she said: "Why can't I pray in the *front?*" At an early age, she had seen the equal conditions granted to men and women in Mecca and Jerusalem. Medina was a wake-up call that showed us the rights that women aren't denied just because they're women. Now, here in Morgantown, she had been the first to lead us through the front doors and into the main hall. She had hauled her heavy baby blue Gap messenger bag, packed with her schoolbooks. She had read about the sit-ins of the American civil rights movement. She was ready for a pray-in.

She wrote her reflections in a notebook. "How would you feel if your own brothers discriminated against you? Well, I don't feel I can go anywhere they are unless I want to come back home all mad because I don't feel welcome."

Unlike many people decades older, she understood clearly what was happening. "Discrimination against feminists is what's happening at our mosque in Morgantown, W.V. Yup, that's right, the majority of these men don't want us praying a foot behind them." As it turned out, they didn't even want us thirty feet behind them.

At that moment one of those men—the Egyptian American board president who had ordered me to use the rear door—descended upon us. His loud voice broke the quiet. "Sister, please! Please leave!" he yelled at me. "It is better for women upstairs." He wanted us in the balcony. When we didn't move, he thundered, "I will close the mosque!" I had no idea at that moment if he would make good on his threat. And I had no idea that our act of disobedience would soon embroil the mosque, and my family, in controversy. Nevertheless, my mind was made up.

"Thank you, brother," I said firmly. "I'm happy praying here."

In fact, for the first time since the start of Ramadan, I was happy in prayer. My father had taken my sleeping infant downstairs so I could concentrate. But my peaceful reverie was interrupted when I heard my father's voice, as the elder went to convince him to tell us to leave the main hall. Words like "discrimination" and "equal rights" filtered upstairs.

That night I returned to the mosque with my father. He slipped into line with the men, who were already starting the nighttime prayer called *isha*.

There were only two rows of about fifty men in the front. I slipped into the vast empty space of the rear of the main hall where I had sat with my family for the dawn prayer. I joined in prayer and sat quietly during the pause. This was the complicated, hourlong *taraweeh* prayer in which congregations recite all of the chapters of the Qur'an over the month of Ramadan. Having never felt welcome at the mosque, and knowing I'd be sequestered in a corner if I did attend, I was thrilled to actually learn the process of this prayer. As it turned out, it wasn't going to be that simple. A man noticed me and whispered to the other men. I heard a voice boom out: "There will be no praying until *she* leaves."

I stayed firmly in position. I knew my rights. Knowledge was power, and I had it. I wasn't going anywhere. The men chased my father out of line and peppered him with pleas to tell me to leave the hall.

"She is doing nothing wrong," my father insisted. I had told him earlier that if anyone wanted to discuss the question of women's presence in the mosque, I was happy to oblige. "If you have an issue, talk to her," my father told them.

Four men bounded toward me. A man from Turkey declared: "Sister, please! We ask you in the spirit of Ramadan, leave. We cannot pray if you are here."

"I have prayed like this from Mecca to Jerusalem. It is legal within Islam," I said. I remained firm.

The Turkish man continued: "But Mecca is a special place. We are weak here."

I wasn't about to take the responsibility imposed on women over the centuries for men's sexuality. I wasn't going to be denied my rights just because I was a woman. "This is your *jihad bil nafs*," I answered, evoking the concept of the holy struggle we must all wage within our *nafs*, or "soul," to transcend our lower selves and rise to our highest spiritual heights.

"You will cause *fitnah*," the man insisted.

I had never heard of this concept of fitnah until I started trying to understand how it is that men rationalize the erosion of women's rights. It means "conflict"; these men argued that women's presence causes fitnah because it is sexually arousing and distracting to men. Fitnah's theological root makes it a very serious allegation. The proverbial anti-Christ, called Dajjal, will emerge in the days before the end of the world and cause strife, or fitnah. Jesus will join forces with a messianic Muslim figure with the title of Mahdi and slay Dajjal with a lance, ending his reign of terror. According to this version, Christ will convert Christians and Jews to Islam, and everyone will live happily ever after, or at least until apocalypse arrives.

I stared at the man. His efforts to intimidate me weren't working.

I knew that I wasn't acting illegally according to the laws of Islam, and I couldn't be forced to leave. I remembered the Qur'an says, "There should be no compulsion in religion."

The man turned to me again and asked, "Just tell me. Are you going to leave? Yes or no?"

I felt fully confident in my rights. "No."

As the Turkish man stood up to leave, an Egyptian student with wide eyes thrust himself in front of me. "Let me just ask you one question!" he yelled. "Is this about you or Allah?"

I knew the answer deep within my heart. This was not about carpet space. This was not about a door. This was a fight for the way Islam expresses itself in the world. September 11 defined the challenge Islam will have in this world during my lifetime. If I was going to raise my son a Muslim, as I had committed to do with his aqeeqa, I, as his mother, had to claim my rightful place within my religion so that I could help to define our Muslim communities by the teachings on compassion, love, and respect that had kept me within Islam's folds. I didn't believe I owed the man an answer to the question of whether my presence in the mosque was about my heart or my ego. I felt that the men would have been better served if they had asked themselves the same question.

The men returned to their lines, and we completed the night's prayers. The thunderstorm had passed without a lightning bolt striking the mosque. But again, it wasn't going to be so simple. The next day the mosque's all-male board met in the mosque kitchen and voted to make the main hall and front door accessible "solely" by men. My father started weeping during the meeting. The men moved into the office. He continued to weep, crying so hard that he started to choke on his tears, losing his breath. He begged the men to reverse their decision.

"Have mercy on me," he pleaded. "My daughter has returned to Islam. Welcome her."

They were untouched. "Everything will be okay for you," they tried to reassure him. It was as if my father's self-preservation was all that mattered. Still crying, he told the board, "Don't adopt this policy. In search of spirituality and peace, my daughter has traveled all over the world, making pilgrimages to the holy places of the world's major religions, including Mecca during the Muslim holy pilgrimage of the hajj. Finally, she has concluded that she will find her home in her own religion, Islam. Treat women with respect," he pleaded. "Give them equal opportunity to pray and to learn about Islam." Alas, his pleas had no effect on the board members. They were solid in their certainty. My father went home and kept the tears he shed at this meeting a secret for months. In subsequent meetings the board members maintained their policy of telling women to pray in a separate space in the balcony.

I continued my act of civil disobedience while having my every request for a formal meeting with the board denied. Inside the main hall I saw the horror of the messages the men delivered to each other about women.

To my shock, at a Friday evening sermon, a visiting imam preached the same message that the book distributed by the Muslim Students' Association had taught allowing "wife beating." He noted that men were to hit their wives with a traditional toothbrush, the *miswak*, made from a root. He talked about this technique as a way to humiliate women. About one hundred men listened dutifully. In the women's balcony, a woman was seething. She had left her husband, sitting down below, because he used this kind of teaching to rationalize his physical abuse of her. Because I was sitting in the main hall, I walked straight to the imam after he ended his sermon and challenged it as inappropriate. He tried to rationalize his points with the miswak argument.

"I'm sorry, but you aren't talking toothbrushes when you use the word *beating*," I declared. I didn't change his mind, but at least I protested. That would never have happened if I'd accepted my third-class status in the balcony.

After one of the final nights of Ramadan, considered a "night of power," my father, one of the men who had started the mosque in Morgantown, gave me an early *eidie*, a gift that elders give on Eid, the festival that marks the end of the holy month. He handed me a copy of the key to the mosque's front door, sold the night before at a fund-raiser. I traced the key's edge with my thumb and put it on my Statue of Liberty key chain, because it is in America that Muslims can truly liberate mosques from cultural traditions that belie Islam's teachings.

"Praise be to Allah," my father told me. "Allah has given you the power to make change." I rattled the keys in front of my son, who reached out for them, and I said to him, "Shibli, we've got the keys to the mosque. We've got the keys to a better world."

ISLAMIC RIGHTS IN WESTERN LAW

MORGANTOWN—The issue of gender discrimination at mosques raises a serious legal question: does the First Amendment of the U.S. Constitution, which guarantees freedom of religion, allow religious organizations to discriminate against women?

I turned to the phone again to find answers. Marcus Owens, a tax attorney at Caplin & Drysdale and former director of the Internal Revenue Service's exempt organizations division, told me, "Muslim women are essentially in the same place as Martin Luther King Jr. in the 1950s and 1960s," when the courts hadn't yet established legal precedents banning racial discrimination.

I was going into a field I didn't understand very well: the law. But it was becoming obvious to me that Muslim men used man-made Islamic law to deny women rights in every facet of life, from their bodies to governance. I had come to the educated conclusion that what these men were doing was not Islamic at all. Western society was not immune from this denial of women's rights in the name of the law. It seemed to me that Western law had matured beyond Muslim communities but not beyond Islamic law, and that Western law and Islamic law were completely consistent in their interpretations of justice, equality, and punishment. Puritan Muslims defined justice with stonings, lashings, and amputations. But those sentences weren't taken from Islamic law. They came from puritan Muslims' *interpretations* of Islamic law. I didn't realize what blasphemy this was to the puritans. But I also didn't care. I wanted to test just how far Western law could go in helping a Muslim woman reclaim her Islamic rights.

What I knew was that most mosques, including the Islamic Center of Morgantown, enjoy the benefits of being designated charitable organizations with tax-exempt status. The Internal Revenue Service has stated that "tax-exempt organizations may jeopardize their exempt status if they engage in illegal activity." Applying the "illegality doctrine" in 1983, the U.S. Supreme Court upheld a lower court ruling that the IRS could deny Bob Jones University tax-exempt status because of racial policies at the evangelical Christian university. Tax attorneys said the ruling established the public policy that tax-exempt organizations can't racially discriminate. The same protection has not been established, however, for gender discrimination; in addition, Bob Jones University is not just a private organization but more specifically an educational institution.

In a 1984 case against the Jaycees, a civic organization, the Supreme Court upheld the ruling that a private organization cannot discriminate based on gender. The case was important for another reason: the Minnesota Department of Human Rights had brought the case against the Jaycees. Even if the federal courts had limited jurisdiction, maybe Muslim women in America could seek recourse through the human rights commissions present in most cities and states.

Attorneys suggested that mosques could also be considered public organizations. The constitution of the Islamic Center of Morgantown addresses "Muslims in the northern West Virginia and other parts of the United States and Canada" and promotes "friendly relations and understanding between Muslims and non-Muslims." A statute of the West Virginia Human Rights Commission establishes a "place of public accommodation" as "any establishment" that "offers its services, goods, facilities or accommodations to the general public." The West Virginia statute doesn't allow an establishment to issue discriminatory rules "directly or indirectly." The statute also prohibits establishments from engaging in "any form of reprisal or otherwise discriminat[ing] against any person because he or she has opposed any practices or acts forbidden" by the state statute. Importantly, the statute protects individuals from being subject to "physical force or violence," "actual or threatened." Like many cities, Morgantown has a human rights commission, with the same mandate as the state's. Although courts tend to stay out of religious affairs, I came to the difficult decision that Muslim women must use the American legal system to restore Islamic rights, particularly when intimidation is used to enforce unfair rules.

Another issue was weighing on me. The national organization with which my mosque was affiliated, giving it tax-exempt status, was the same organization that had given me such a rich experience on the hajj—the Islamic Society of North America. My mosque's constitution said that the mosque couldn't do anything that violates Islamic standards; if it did, the society could send a representative to mediate. And then there was a national organization established as a sort of ACLU for the American Muslim world. I had first heard about it earlier in the year when I tried to help an Iranian graduate student who'd been detained in an FBI investigation. He ended up being deported. I learned from working on his case that the Council on American-Islamic Relations was set up in Washington to protect Muslims' civil liberties. I was a Muslim. My civil liberties were being violated through gender discrimination at the mosque.

The issue of women's civil liberties had been very much a taboo topic in the Muslim community. Even though the civil liberties group and other organizations knew about systemic gender discrimination at mosques, having done the survey that documented it, they hadn't done much to remedy the problem. The civil liberties group told me that it had never received a complaint from a Muslim woman citing discrimination at her mosque. I sat at my computer late into the evening and diligently wrote out the first complaint filed with the group for gender discrimination at a mosque. I alleged that my mosque's board imposed separate and unequal accommodations.

I wondered what would happen.

TIME FOR ACTION

MORGANTOWN—Halfway across the world, a band of strong-willed Muslim women in a village in India were giving me strength. An e-mail popped up on my computer with the headline: "Women in Indian Village Fed Up with Sexism, Build Their Own Mosque."

The *Hindustan Times*, a leading daily in India, reported that in the village of Parambu in the state of Tamil Nadu in India's south, not only were women banned from the local mosque, but village men expected them to accept divorce rulings doled out by a male-only religious tribunal that met, of course, at the off-limits mosque. Through a fledgling organization called Chaaya, or "Shadow" in Sanskrit and Hindi, the women had decided to start their own mosque to educate women about their legal and Islamic rights. They were led by a woman identified only as Sharifa. In Arabic, Sharifa means "honorable." I sent Dr. Mattson the article about the women in Tamil Nadu and a note: "Here are some Muslim women who know their rights!"

The courage of Sharifa and the women in Tamil Nadu meant so much to me, comfortable as I was in my life in America. They were women who were really taking a risk to defy tradition to defend their rights. They underscored for me that the battle over women's access to the Morgantown mosque wasn't just a fight over carpet space. It was emblematic of the systemic marginalization of women in Muslim communities. Even in Morgantown I had been able to see that the men used the mosque as a watering hole where they gathered to contemplate and reinforce decisions that intrinsically deny women their rights. For example, the man at our mosque who beat his wife got support from other men when his wife appropriately left him.

The women in Tamil Nadu reinforced for me that I had to exercise my talents to challenge the barriers to entry that women faced in their local mosques. I tapped at my keyboard until I had an essay. I pushed the SEND button, delivering it to the *Washington Post*. My editor there, a talented woman by the name of Kathleen Cahill, worked with me over the next days so that I could lay out the most convincing argument Islam offers for women's rights in mosques. To test the waters, I sent a draft to Saleemah Abdul-Ghafur, a young Muslim woman I was getting to know. She was one of the founders of a Muslim women's magazine, *Azizah*. A talented friend of mine, Kemba Dunham, a reporter at the *Wall Street Journal*, had introduced me to her earlier in the year after my first book came out, thinking Saleemah would appreciate the book. Saleemah wrote back quickly with an expression of support that I very much needed. "I acknowledge your courage for writing such a thoughtful and bold piece. I too have been *very* frustrated with the inequities that immigrant and indigenous Muslims practice in America. Indeed some take pride in their oppression. I am so glad that you didn't do as I and many others have done and just stopped attending mosques. You acted against wrong—the strongest expression there can be of faith." She ended with a salutation that I appreciated: "In solidarity." And she made a point I little understood: "Asra, you're startin' a revolution."

On December 28, 2003, the *Washington Post* published my essay under the headline "Rebel in the Mosque: Going Where I Know I Belong." Inside was an illustration of our morning ascension, drawn by Samir, the new Mussalman. It showed the vast space between the men and us.

I had rarely confided my personal sense of disenfranchisement to fellow Muslims for fear of backlash. Outside of the hajj, I found that the community could be mean-spirited. Deciding to take the risk, I agreed to have the essay posted on an unconventional Muslim website I'd recently discovered, Muslim WakeUp! The reaction was immediate. Muslim women and men who shared a vision of equitable mosques responded swiftly with electronic messages of overwhelming support. "Beautiful, simply beautiful!" read the first remark posted by a woman named Maria on a comment page.

This website had empowered me through the voices it publishes, including a powerful poet named Mohja Kahf, a Syrian American literature professor at the University of Arkansas in Fayetteville. She made me laugh and cry every time I read a poem she wrote, "My Little Mosque," about the crack addicts and Dumpsters that women have to pass before getting to the women's entrance at a mosque, that, while fictional, captures so many of our experiences. She wrote to congratulate me for speaking out. "The revolution has begun!" Her words meant the world to me. And other e-mails kept pouring in: "This is a powerful article and it has inspired me with hope and determination. I hope the author continues with her certitude and convictions and that many more sisters will join her. We need Muslim sisters who aren't scared to use their own intellects and wits in combating the extreme patriarchy which sees the woman as *awrah* [denoting forbidden] and *fitnah* [denoting conflict]," wrote a Muslim man named Adam Misbah al-Haqq. In the first month, about three hundred people e-mailed their thoughts directly to me or published their views in various Internet venues.

To analyze the responses, my father and I joined forces in a beautiful way. We co-authored an article for the *Journal of Islamic Law and Society*, "Gender Apartheid in Mosques: The Divide Between Men and Women." My father did the statistical analysis on the letters I'd received and the messages that had been written about the article.

About nine out of ten Muslims, or 91 percent of 205 respondents, supported improved rights for women in mosques. Just about all non-Muslims, or 99 percent of 108 respondents, expressed support. My father offered his special scientific endorsement of the results: the difference between the two groups was statistically significant, and the same findings could be extrapolated for larger populations. Men were slightly less supportive than women. All non-Muslim women expressed support, and almost all non-Muslim men expressed support. A non-Muslim woman sent a one-line e-mail: "Be strong—people like you are the salvation of Islam."

The backlash caused me restless nights. Men expressed more judgmental attitudes than women. The judgmental reactions focused in large part on fifty-five words I had included in my fifteen-hundred-word essay to explain the intimate context of my relationship with Islam. "When I became pregnant last year while unmarried," I had written, "I struggled with the edicts of some Muslims who condemned women to be stoned to death for having babies out of wedlock. I wrote in the *Washington Post* about such judgments being un-Islamic, and my faith was buoyed by the many Muslims who rallied to my side."

About 20 percent of Muslim men were judgmental on either the issue of my unwed motherhood or the issue of public discussion of internal issues, compared to 10 percent of Muslim women who were judgmental. A Muslim WakeUp! reader who wrote that I should have put Shibli up for adoption by a two-parent household rather than raise him as a single parent later rescinded that suggestion and publicly apologized to me for it. In a Yahoo discussion group for Indian Muslims, a Muslim man wrote: "Asra Nomani is the latest publicity speaker/profiteer in the same string as Salman Rushdi and Taslima Nasreen. She is simply trying to malign the Muslim community in America." Supporters of improved women's rights were critical on these two points. A Muslim man from Seattle wrote expressing support for sharing prayer space with women, but noted: "Never have I read, heard or seen any evidence that 'it is o.k. to have a bastard' and brag about it. . . . May *Allah* show us [all] the righteous path, and forgive our sins. *Amin* [Let it be so]!!"

I sent my new friend Saleemah one of the e-mails from a critic, a Muslim man who wrote to ask whether I was "ashamed" to admit my unwed motherhood. "We don't need people like you to comment on Islam. Do you have any idea what is right and what is wrong in Islam? It seems that your father is also a shameless person like you. We don't need this type of new Islam. Either change yourself or be out of it," he wrote from the e-mail address "onlythe_truth2003." Saleemah sliced his arrogance with humor. She wrote back, "His email address should be 'onlyhis_truth1803,'" recognizing that his attitude was akin to Western cultural attitudes from the nineteenth century.

There was clear evidence of systematic discrimination against women around the country. Of all the responses, sixty-four women and men mentioned incidents of discrimination at mosques. A sixty-three-year-old convert to Islam said she regularly confronted men and women who wanted to move screens in front of her and other women during prayers and meetings. Other women wrote about hostile environments toward women at their mosques. Six women cited fair treatment at mosques.

The worst fears of some Muslims that critics of Islam would seize on the issue to demonize Islam didn't materialize. Only about one in ten non-Muslims negatively judged Islam because of the public disclosure of inequity at mosques. Instead, readers and the media made two important distinctions: they separated cultural traditions from Islam, and they recognized gender discrimination as a universal phenomenon. Five Christians cited discrimination against women at churches; four Jewish respondents cited discrimination against women at Orthodox Jewish synagogues.

In a letter to the editor of the *Dominion Post,* the Morgantown newspaper, that was headlined "Scarves Off to Nomani for Taking a Stand," Morgantown photographer Sue Amos wrote: "Nothing like shining a little light on the darkness. . . . And we all know sex discrimination exists in most religions." She cited a wedding she had photographed in a Baptist church where the minister declared, "Women are the weaker vessel and must submit to their husbands," among guests whom the author knew included battered housewives. "No more back doors, Asra," Amos wrote.

The *Charleston Gazette* ran an editorial, "Woman's Place: Taking a Stand for Equality," arguing:

> This issue encourages everyone to do some investigating, to discern the difference between cultural expectations and the real demands of a faith, whether it is in Islam, Christianity or Judaism. All faiths are susceptible and, at different times in different places, have been bent to serve a purpose that has little to do with their adherents' relationship to the eternal. All faiths at their best treat each of their faithful with dignity and make no demand that requires them to surrender it.

Amir Kanji, the Ismaili businessman from California who had invited me to the Ismaili health care professionals' conference in Atlanta, wrote to me: "It is a tragedy that we perpetuate cultural mores in the name of Islam particularly in America and that in the year of 2003." His life continued to express Islam's values of inclusion, tolerance, and respect. His eldest daughter had given birth to a daughter three weeks prematurely on what would have been his father's ninetieth birthday. Her parents gave her a name, Lyla Sofia Miller, that reflected the passing of Islam into a uniquely American Muslim identity. Amir planned on celebrating Thanksgiving in New York with his new granddaughter in his arms.

It became clear to me that it is a victory for Islam if Muslim men and women establish the rights of women to full participation in mosques in America. A combination of community organizing, education efforts, legal complaints, and cultural sensitivity training will be necessary to transform mosques from men's clubs into centers that welcome everyone—men, women, and children. There is clearly a groundswell of support among ordinary Muslim men and women around the world for such an effort. Acknowledging the injustice prompted me and others to devise a campaign to educate Muslim women and men about women's rights in the mosque. I started dreaming up a campaign that Dr. Godlas suggested we call "Take Back Your Mosque," and I started drafting a "bill of rights" for women at mosques. I had the support of my new friends in Islam.

I talked to the editor of Muslim WakeUp!, an Egyptian American who had been raised in southern California, Ahmed Nassef. He was creative, visionary, and kind, having launched a "Hug a Jew" program. Born in Egypt, he moved to Los Angeles at age ten, majoring in Islamic studies at UCLA and working in marketing in New York and the Mideast before starting his unconventional website in 2003 to represent the "vanguard of the progressive Muslim movement" for social justice, gender equality, pluralism, and free inquiry into the full range of the religion's 1,400-year-old traditions. "It's time for action," he told me. "The days of women being relegated to the attic and being forced to use the back entrance by the Dumpster are numbered."

A RESTRAINING ORDER IN THE HOUSE OF GOD

MORGANTOWN—On a Friday night in January 2003, I got ready to go to the mosque in Morgantown for the weekly halaqa that only men seemed to attend. It was a different story in Mecca, where, on the hajj, other women and I attended and participated in all of the same study sessions that men enjoyed. There was no segregation, separation, or even discouragement in Mecca. We sat wherever we wanted to sit. Sometimes I sat right next to my father or other men, as did other women. There was no discomfort. I asked questions freely, as did other women. It was the same at the national convention of the Islamic Society of North America, where men and women had sat freely wherever they wanted and were not excluded from any events.

It wasn't right that in Morgantown men enjoyed weekly study sessions, often along with free dinners, and women were not welcome. I wanted to know what they were learning. But I was also afraid. I wrote to Saleemah, "How sad. I am about to go to the mosque, and again I face my own fears. I am going to the weekly Friday night *halaqa*. Women do not go. I want to partake in this religious education the brothers of my community receive. But I am afraid of the reception I will receive. But I will have with me all the sisters and brothers with whom you have connected me. Thank you." Saleemah wrote back immediately: "Yes, take us with you. Call on the powerful, free, and fully self-expressed woman that Allah made you! And remember the legacy of Hajar, Khadijah, and Ayesha, may Allah be pleased with them. In solidarity, Saleemah."

I closed the e-mail, internalized its message, and set off for the mosque with my father. There I walked inside with parts 1 and 2 of *The Meaning of the Glorious Qur'an,* with commentary by the Islamic scholar Yusuf Ali. His translations were considered among the most lyrical. I slipped to the floor at the rear of the main hall, about thirty feet behind a huddle of men reading the Qur'an to each other. I closed my eyes and started meditating. I saw all the beautiful people of Islam whom I had gotten to know since I had gone on the hajj.

There were the noble brothers in Islam who had been supportive and kind: Dr. Godlas, Ahmed Nassef, Dr. Omid Safi, Michael Wolfe, and Alex Kronemer. There was the Chicago lawyer-comedian, Azhar Usman, at the Islamic Society of North America conference who made us laugh at ourselves as a community divided over how to properly greet each other. There were the sisters in Islam, strong, feisty, and expressed: a rap poetess in hijab, strong and determined at the ISNA convention, where leaders recognized they couldn't silence the women and youth; Ingrid Mattson, the vice president at ISNA who acknowledged that mosques are men's clubs; Saleemah; Mohja.

Hearing noise around me, I opened my eyes. In the front of the room a Syrian American cardiologist from Clarksburg was taking a seat. He was a member of the board and had voted to make the main hall and front door "solely" for the use of men. The halaqa was about to begin. I remembered the study sessions in Mecca and Medina. Men and women sat in the same space, close to the speaker, and I had asked questions without fear or difficulty. In Morgantown I moved forward to take a seat behind the men, sitting at least ten feet behind them.

"Sister, sit in the back," the cardiologist ordered.

I didn't know if I had heard right. Surely, he had greeted me in the salutation typically exchanged. I answered him as if he did. My father later confirmed he had not. "Walaikum as salaam, brother," I said, giving the reply that usually comes after a greeting. I tried to respond to his demand. "What are you suggesting? That I sit in the back against the wall?" I said, pointing about thirty feet behind me to the dark back wall where rolls of unused carpet were tossed.

"Yes."

I couldn't understand. I had sat not even behind men but next to men to listen to lectures during the hajj in Mecca and Medina—the two holiest cities of the Muslim world. Nobody had demanded that I sit even ten feet away from the men. Even the conservative Sheikh Alshareef, our hajj tour guide, looked me in the eye when he answered my questions. But somehow in Morgantown, this man was trying to turn me into a pariah. I refused to be cast aside. "Brother, I have an Islamic right to sit here and participate."

He snapped at me: "The brothers do not want you here."

"How many brothers want me to sit in the extreme back?"

The group included professors and tomorrow's leaders—graduate students, engineering students, and medical residents. Slowly each hand but my father's hand rose.

In my book, the majority didn't define justice. If that had been the case, Islam would never have survived the rejection that the prophet Muhammad initially received in Mecca. America would never have integrated if the issue had been put to a vote in the white South. "It is not Islamic to impose these conditions on women," I said.

"The mosque board made rules. Your father is on the board," he answered. This was the first public admission of the ban that the board passed the day after my mother, niece, and I walked through the front door to pray in the main hall. They hadn't yet been publicly posted or expressed.

"My father did not agree to these rules," I said.

My father, otherwise quietly observing until then, interceded. "These rules are wrong."

"I am the teacher. Why do you come to listen to me if you cannot respect me?" the cardiologist asked me.

"But, brother, you are not respecting me."

A man turned toward me with anger in his eyes. He was an Egyptian research assistant professor of engineering at WVU. "Leave," he yelled at me, as if he had a right to make such an order. He continued his tirade, thrice calling my father "idiot" and flailing his arms so wildly at my father that a man from Turkey had to restrain him.

When I protested his disrespect, he said, "He is an idiot. Look at the kind of daughter he raised."

This was a displacement of blame with which I was familiar. It was the same mentality that blamed a woman for distracting men. Then the cardiologist threatened me with a restraining order, banning me from the mosque. When I refused to budge, he cancelled the halaqa, but he just moved to the other side of the hall, the men, including my father, encircling him and shutting me out. I didn't resent my father sitting with the men. It was a confusing moment in which tradition and justice clashed.

I realized there were wider dynamics at play. In my effort to do social service after I'd returned from the hajj, I'd become a volunteer for our local Rape and Domestic Violence Information Center, which ran a shelter and hotline. In my training, I learned about something called the Power and Control Wheel. It was a model for what abusers commonly eked out to those who challenged their authority in domestic violence situations. I could now see that it applied to all aspects of life, including my experience with the men in power at the mosque.

I closed my eyes to meditate. In the midst of troubles, the prophet Muhammad had gone within himself to find a divine answer. I could at least seek out human insight. In the course of one hour, I had gone from such appreciation for the *sangham*, the spiritual community, I had found in Islam to despair. I could feel my soul quivering from the way these men had physically threatened my father and me.

When I rode my Hero Honda Splendor motorcycle through India, I meditated on the image of the fierce Hindu goddess Durga on a tiger, not because I worshiped her, not because I was committing *shirk*, or idolatry, but because this image captured the spirit of a strong woman. In mythology, she slays fears. At that moment in the mosque, the darkness of my lids shielded me from the violent men around me. I went back to a graveyard in Mecca and saw dear Khadijah, the first wife of the prophet, commanding her caravans, challenging her husband. I saw Aisha, the youngest wife of the prophet, in the mosque, chronicling Islamic history and theology. I saw dear professor Ingrid Mattson, sitting poised and dignified with her intellect and power on the dais of the Islamic Society of North America convention. I saw my mother, my anchor. I saw Safiyyah, the future, our hope.

I knew that people often lose sight of basic human decency. When we were returning from the hajj in Amman, Jordan, we watched a pack of irate pilgrims stampede the gates to storm our plane so that they could get seats on it. They had missed their own connection. Yelling and screaming, they delayed our departure so long that we missed our connection from New York City home to Morgantown. I wondered how we had become so indecent as an ummah. A single tear sat locked between my eyelids.

DISAPPOINTMENT AFTER DISAPPOINTMENT

Wake up from your sleep and say, "Bismillah [In the name of God]."
Dawud Wharnsby Ali
(Canadian Muslim singer)

MORGANTOWN—The next morning I awakened before dawn. I had forced myself to go to sleep, but my night hadn't been restful. I knew I was physically safe. A woman dispatcher for the Morgantown police told me that the state of West Virginia didn't issue restraining orders in nonfamily disputes. But I was spiritually brutalized.

I slipped a CD into my stereo system. The singer was Dawud Wharnsby Ali, a Muslim convert living in Canada. A hajji, he celebrated the hajj with a song, "Here We Come," translating the prayer we had chanted going to the Ka'bah and to our mosque in Morgantown the morning of our predawn ascension. I'd bought the CD at a table at the bazaar at the Islamic Society of North America convention. I had seen a familiar face behind the table: Suhaib al-Barzinji, a bespectacled man. He was one of the guides during the hajj, and he had helped unload our suitcases when our bus pulled into Mina. He was running a booth for Astrolabe, the Islamic media distribution company he ran with his wife. I was walking the floor of the bazaar at the time with Sheikh Alshareef, my group's guide around the Ka'bah and elsewhere. Unlike the men at the mosque the night before, the men attending the convention talked respectfully and freely with me. Nevertheless, the horror of the night before stayed with me. I had wanted to write about the beauty I had witnessed in the Muslim community. Instead, I found myself writing through a flood of tears in the stillness of the dark predawn. I sobbed so hard that my body shook. Dawud Wharnsby Ali's singing spoke to me: "Wake up from your sleep and say, 'Bismillah.'" In the name of God. The word was meant to be a touchstone for staying mindful of a higher calling in all of our actions.

When would Muslims awaken as a people? I had witnessed so much need for awakening. Erum Afsar, a young woman in Canada, wrote to tell me about the neighborhood mosque she was afraid to enter because of its hostile policy toward women. Pamela Taylor, a Muslim convert and writer, had to step on a path of crushed, rotten cherries leading to the rear women's entrance at a mosque in south Florida; Sara Tariq, a young Arkansas physician, lost access to her childhood mosque in Little Rock when conservative men took over. I thought I would feel triumph for the ummah that at least we had women and men who were testifying about injustice. Instead, I was weeping.

The songwriter's words echoed in my house with the fears he imagined the prophet Muhammad felt on his deathbed for the future of the Muslim community, the ummah. He sang of the stillness around the prophet as the companions gathered near his deathbed. The prophet died on June 8, 632, at the age of sixty-three, just one day after the date that marks my birthday. "As Aisha, his wife, held tight to his hand, the prophet spoke again before he passed away. My ummah, those who follow me, the future of the faith makes me worry until I cry. My brothers and my sisters in Islam, will they be strong and carry on after I die?" I asked myself these questions through my tears: Brothers and sisters in Islam, will we be strong? What is the future of our faith?

That afternoon, in hope, I turned again to the Council on American-Islamic Relations. As a Muslim woman, I was being physically and verbally threatened for not leaving my mosque. I pulled out the mobile phone number for Ibrahim Hooper, one of the organization's founders and its spokesman, and called him.

"As-salaam alaykum," I said, hearing shouts behind him.

"Walaikum as salaam," he answered. "Could you wait one minute?" he asked. Behind him I could hear a man shouting, "Allahu Akbar," prompting Ibrahim to repeat the call, "Allahu Akbar!" God is great!

He was in front of the French embassy in Washington, D.C., taking part in the protests that Muslims were holding nationwide against the ban on head scarves in public schools in France. The protesters cited article 18 of the Universal Declaration of Human Rights: "Everyone has the right to freedom of thought, conscience and religion; this right includes freedom to change his religion or belief, and freedom, either alone or in community with others and in public or private, to manifest his religion or belief in teaching, practice, worship and observance." The Muslim world was galvanized to protect the rights of women to wear scarves on their heads, but I was left standing alone in my mosque in Morgantown as the men attacked me for trying to exercise my right to observe my religion in "community with others" and seek knowledge. The irony struck me. At a time when Islamic organizations and civil rights groups were pumping resources and organizing tens of thousands of Muslims from Alabama to France to sign petitions, hold rallies, and uphold the right of women to wear hijab, Muslim women were being denied fundamental Islamic and human rights to speak, participate, and learn in mosques throughout the United States and Canada. Men could feel noble protecting our right to wear a cloth over our hair, but they went silent when it came time to protect our right to speak.

Ibrahim Hooper returned to the phone: "Yes?"

"You're working so hard to defend Muslim women's rights to wear a scarf over their head. More power to you, but meanwhile Muslim women in America can't even exercise their rights in mosques." I was sick to my stomach as I explained the events of the night before, shocked at how my voice was trembling. The hostility I'd met with had disturbed me more than I dared to admit to myself. He listened to me and then said, "You're in the wrong town." I was aghast. "Is that what you would say to a Muslim woman who is disrespected at her place of work in any town in America?"

He said he would look into it and got off the phone.

It was obvious that the community would not rally to protect women such as myself. I told my father I was going to report the incident to the police. I didn't ask him to go with me. He volunteered. That meant so much to me, because often we are not just fighting the status quo when we try to challenge authority but facing off against our loved ones as well. I drove my father and myself to the Morgantown Police Department, a sprawling building in downtown Morgantown just two blocks from where my father had started the mosque and my mother had started her boutique. My voice trembling just slightly, I related the incident to a policeman, Officer M. J. Bloniarz, who listened intently. "Did you feel physically threatened?" he asked. "Yes," I said. A local, he told me that not long before he had been called to an accident involving a Muslim woman who was wearing the full face-covering veil popular in Saudi Arabia. Her husband, in the passenger seat, wouldn't allow her to answer questions from the police officer. "It is my religion," he told the officer. "What did you do?" I asked. He let them go. "I didn't want to be disrespectful." I shook my head in amazement. That incident was yet another example, even minor, of how the puritans were wrongly using Islam to barricade themselves. After all, the woman's mahram, or sanctioned male relative, was present, and the conversation would have been not only chaperoned but purely business.

I completed the paperwork and became incident report number 04-2646. The police classification: assault intimidation. I felt better for having documented the spirit of intimidation with which the puritans were trying to keep their hold on not only our mosques but the Muslim world.

"Do you feel okay about it?" I asked my father as we drove away. "I have no regrets," he said firmly. "The extremists cannot just bully their version of Islam upon everyone."

I sent the report to Saleemah the next day on Martin Luther King's birthday, "with great sadness about the present but hope for the future." "Oh, yes," she wrote back. "On this day that we celebrate the legacy of Dr. King, I can't help but remember his quotation, 'We as a nation must undergo a radical revolution of values.'" He spoke those words in 1967 to oppose the Vietnam War, but his words were universal in the struggle of all people to act on their conscience." She ended: "Onward."

In a field trip to the Pittsburgh Zoo with our families, the local Muslim woman who courageously left her husband after he constantly beat her taught Shibli to drink from a straw for the first time. That afternoon, as we raced from the elephant house to the monkey house to get out of the cold, I turned to my mother to make sense of the men at the mosque. "Why are they so mean?" I asked. "Power and control," she answered. "Asra, it is not easy for anyone who wants to bring about change. They hated Jesus. They hated the prophet Muhammad."

Her thoughts made me remember a moment in Mecca when we saw the site of a house next door to the prophet's house where Abu Lahab, his uncle, had lived, tormenting him. Interestingly, the spot was now the site of a men's restroom. Samir had studied the spot and asked, "His evil uncle lived next to him?" That's right, I said.

When the prophet and his army defeated the soldiers of the Meccan Quraysh tribe in the Battle of Badr, it's said, Abu Lahab, in a fit of anger, struck a servant who sympathized with the Muslims. A woman, Ummul Fadhl, or "mother of Fadhl," hit Abu Lahab on the head to rebuke him and succeeded in silencing him. I drifted to sleep that night with this aimless thought, praying that somewhere, in the course of the long, dark night, somebody could smack some sense into the minds of those who op-posed change just to hold on to power and control.

I found some hope. The secretary general of the Islamic Society of North America called the cardiologist to tell him the mosque needed to back down from its exclusionary policy toward women. The research as-sistant professor apologized to my father, who accepted his apology. And some days later, on a Friday night, I opened an e-mail message from the acting president of the mosque. He thanked me for my "input" into the mosque's operations and said, "For your information Islamic Center of Morgantown has a policy that every Muslim man or woman is encouraged to pray and is welcome to attend masjid services."

He noted in a paragraph set off from the rest of the letter: "Sister, the Board of Trustees wants to assure you that no discriminatory policy will be implemented in the Islamic Center of Morgantown."

"A victory!" I shouted to my mother, bounding up the stairs with a printout to show her in the kitchen. She studied it and said, "In three months, you've done what your father couldn't do in thirty years."

My father smiled proudly and hugged me. "Al-hamdulillah!" Praise be to God! But after the buzz of the victory was over, he, the son of a defense lawyer, sent me back to the computer to clarify that the statement af-firmed the rights of women to the front door and the main sanctuary, in-cluding all activities there, such as the study sessions. I got no response to this query.

In separate personal conversations, about six Morgantown Muslims blamed me for tarnishing the image of Islam and the Morgantown Muslim community. A Muslim man sent me an e-mail: "As your eternal brother in Islam I strongly support your ends; but as a member of this community and of the greater *ummah* I vehemently disagree with your means."

I met him, a graduate of the West Virginia University School of Law, at a local coffee shop, Panera's Bread. He had praised my *Washington Post* essay in an e-mail to his father, a former West Virginia University profes-sor who was in Egypt. But then he talked to the conservative men at the mosque. I had shamed the mosque, he told me.

"Stop writing," he said.

"I won't," I said. But I would be happy to meet with the men at the mosque, I told him, to discuss how they could bring about changes that would make the community more welcoming to women. He told me he'd call me back about setting up a meeting. He never did.

IMPORTED HATRED

MORGANTOWN—Sitting in the safety of my hometown, I had to admit I'd gone to Saudi Arabia with a bias against the country. I didn't subscribe to the rhetoric of division that I had heard was promulgated by its ranks of mullah, or religious clerics. I returned from Saudi Arabia, however, having heard only hints of disturbing rhetoric and without having witnessed blatant expressions of hatred toward people who didn't agree on theological doctrine. The Saudi government was starting to wise up to the dangers fomented by hate speech.

That was why I was stunned when I sat one Friday spring afternoon in my corner of the mosque, a figurative bubble around my sister-in-law and me. My mother, who had somehow known he shouldn't listen to what was about to come from the pulpit, had gone downstairs with Shibli. A short, bearded graduate student in engineering from Saudi Arabia stood at the new *minbar*, or pulpit, constructed for the delivery of *khutbah*—the Friday sermons. Until I had started going into the main hall, I didn't have a clue what that word meant. If I had heard it a year earlier, I would have thought someone was clearing her throat. But after listening to sermon after sermon, I'd come to understand not only its literal meaning but also its political significance. In his research, my former professor at American University, Hamid Mowlana, had identified the ways in which the Friday sermon has been used as a political tool by clerics to rally congregations. For this reason, after Pakistan became an ally of the United States in its hunt for Osama bin Laden following the 9/11 attacks, the government tried with only limited success to crack down on clerics at Pakistani mosques who were urging men to join the Taliban in its jihad against the United States.

As the words started spilling from the Saudi student's mouth, I started to wonder where I was actually sitting. "To love the prophet is to hate those who hate him and the Sunnah," he declared emphatically.

"What?" I mouthed to my sister-in-law. She looked at me, wondering, I could tell, what I was going to do. This was not the message my parents had taught me at home. I started scribbling his words into my notebook. Even in Saudi Arabia the leaders knew the line to toe. King Fahd and Crown Prince Abdullah had sent a message to pilgrims during our pilgrimage: "We have to build confidence among ourselves and with other nations by adhering to the teachings of Islam, which rejects isolationism and prevents the seeds of hatred being sown among people." They preached the message that my parents had taught me since my earliest days. "Islam does not prevent its followers from dealing with people of other faiths," they said. "It is a religion of tolerance and calls for peaceful coexistence with other communities."

Over and over again, I had heard that the word *islam* is an etymological cousin to *salaam*, the Arabic word for peace. But in Morgantown this student was railing against man-made laws that have replaced sharia and blasting the "contemporary enemies of Islam" who do not adhere strictly to Sunnah, calling them "evil" and "wicked." It was the classic rhetoric of Wahhabi and Salafi ideologies, both of which have been used, in part, to breed militant sectarian attitudes against non-Muslims and Muslims who disagree with them. As far as I could tell, they were *not* espousing violence. But they *were* on that slippery slope of dogmatism and intolerance that is extremely dangerous to democratic society. American Muslims may live in a country that has not been the paragon of tolerance with its history of racism, sexism, and, most recently, civil liberties abuses in the name of the Patriot Act, but it is vitally critical that they nevertheless rise to the highest principles of America's and Islam's benevolent teachings.

From the beginning, I knew my battle at the mosque wasn't just a women's issue. I believed that intolerance toward women serves as a predictor of intolerance toward others. Amy Leigh, the mother of Shibli's first Morgantown playmate, a boy named Alex, had given me an article in a magazine put out by her church, the Unitarian-Universalist Church. The common denominator in all fundamentalist religions, it said, is sexism. Another ingredient: intolerance toward people of other faiths. Another friend gave it to me in simple words: making women invisible is a main ingredient of violent societies.

In my mosque, what was alarming was not only that this man, living with two Saudi wives in Morgantown, spewed this hate-filled rhetoric just blocks from the campus of West Virginia University, but that not one of the 150 or so WVU doctors, professors, professionals, PhD students, and undergraduate students in the congregation uttered a protest. From the trenches in small-town America, I was observing something disturbing. Even at a time when the government of Saudi Arabia was taking a more moderate position—at least publicly—tolerant and inclusive Islam was losing in places like Morgantown as zealots filled a vacuum created by an ambivalent moderate majority and a passive, even sympathetic, leadership.

My dear friend Pam was just back from London. The British police had arrested a band of young Muslim men for alleged terrorist activities. She said that what was so unusual was that the Muslim leaders had taken to the airwaves to encourage Muslims to help the British in their efforts to keep the society safe. The Blair government had gone to these leaders and said that it would work to protect the civil liberties of Muslims if they would help law enforcement eliminate terrorist threats. Just two days before the sermon of hate in our little mosque, the Muslim Council of Britain had sent a letter to one thousand British mosques urging their members to cooperate with antiterrorism efforts, to oppose extremism, and to provide "Islamic guidance" to help "maintain the peace and security of our country." I saw our choice clearly. American Muslims needed to follow the lead of the mainstream Muslim Council of Britain and acknowledge and eliminate the intolerance, zealotry, and hate that quietly permeate so many of our communities.

About one in four of America's mosques is affiliated with the Islamic Society of North America, and most, like mine, have enjoyed tax-exempt status because of that affiliation. The Islamic Society needs to send a clear message to mosques and appoint a representative to respond to acts of intolerance and bigotry in America's mosques so that moderates will have some recourse other than simply abandoning their mosques to zealots. The society is implementing a long-term solution with new leadership training, but the future of Islam in America also urgently needs a short-term policy on the rhetoric of hatred and intolerance. Infuriated at what they consider to be meddling in Muslim affairs, even liberal Muslims are mocking a plan by the controversial conservative Daniel Pipes to set up an organization under the banner of "progressive Islam" to counter extremism. But instead of being infuriated, we need to stand up from within. I listened to CIA director George Tenet testify to the 9/11 Commission. Borrowing from his assessment of his agency, we need revolutionary—not evolutionary—change in the culture of our mosques and communities.

My little mosque had become a caricature of what happens when Muslims don't take action. A band of men had staged a coup, not only symbolically seizing the power of the pulpit but one Friday literally unveiling the looming new pulpit from where their men preached in the long beards and short robes characteristic of the firebrand school of Islam they followed. The coup leaders included the Saudi student who preached hate, an Egyptian mechanical engineering student who expressed support at one Friday sermon for Hamas spiritual leader Sheikh Ahmed Yassin, and a graduate student in engineering who screamed at us in another Friday sermon not to mingle with the *kafir*, or nonbelievers. The coup leaders included a cadre of students and three professors from the WVU Lane Department of Computer Science and Electrical Engineering, which has research contracts with the U.S. Energy Department and the National Science Foundation. One of the grants includes dental forensics work in collaboration with the FBI; another involves computer security work for NASA. In a sermon, the leader of the takeover, a WVU professor of engineering, alluded to the "immoral" people of the West who are on a "dark path," telling us we should neither mingle with Jews or Christians nor read their holy books.

I went back to the guidebook I'd read from when I walked in Hajar's footsteps between Safa and Marwah in Mecca. I looked up the author, Sheikh Al-Uthaimin. He was another Wahhabi cleric, also allied with the Salafi ideology. I found him listed on a website called Salafi Dawah Online. *Dawah* means to teach Islam to others, most often non-Muslims. I read an account of another book he had written, *The Muslim's Belief*. It had a disturbing passage: "It is our opinion that whoever claims the acceptability of any existing religion today—other than Islam—such as Judaism, Christianity and so forth, is a non-believer. He should be asked to repent; if he does not, he must be killed as an apostate because he is rejecting the Qur'an."

The murder of my friend Danny Pearl had made all of this hate talk unacceptable to me. I remembered a moment on our pilgrimage when my family and I were driving to the cave outside Mecca where the prophet Muhammad had received his revelations. Safiyyah, Samir, and I had been talking about Danny. "Why did they kill him?" Safiyyah asked gently as we drifted past a Sunoco Service Center. A bus of African pilgrims had departed as we arrived at the base of the cave, near a building marked Turkish Snacks.

"I don't know for sure, Safiyyah, but they hated Jews. And Danny was Jewish. They made him say, 'I am a Jew,' before he died, as if that was reason enough to kill him." Samir thought about what I said and gave me the wisdom of a child: "Why would they hate him if Allah put him on this earth?"

As I pondered the frightening rhetoric spilling from the pulpit at my local mosque in Morgantown, I knew my God didn't want us to hate each other. I protested the sermons to our local mosque leaders, the Islamic Society of North America, and even the Council on American-Islamic Relations (CAIR). The Islamic Society told me it wouldn't interfere in local matters. To my delight, CAIR had started a campaign, "Hate Hurts America," to counter hate speech. I documented the hate speech at my mosque, but to my disappointment the group meant hate speech on conservative talk radio, not at the mosque pulpit. When CAIR founder

Ibrahim Hooper told me, "There is a difference with hate speech at your local mosque and talk radio that reaches millions," I answered: "But you would protest, appropriately, if one sixth-grade teacher anywhere trashed Muslims." He said, "Send me the material again." I did but never heard back.

MESSAGES OF WORTHLESSNESS

MORGANTOWN—At the weekly Friday sermon at our mosque, a graduate engineering student from Egypt stood in front of our congregation of about 150 doctors, professors, professionals, and students and proclaimed: "A woman's honor lies in her chastity and modesty. When she loses this, she is worthless."

"Worthless?" I mouthed to my mother.

Worthless, I repeated in my mind. By this man's judgment, that would be me. My empty wedding ring finger and my son, at that moment tossing sand out of a sandbox at his friend Alex's house, were evidence that I was an unchaste woman. It little mattered that I accepted the responsibility of my son's conception to love him fully and completely, while his father abdicated his responsibility. I was confused. I was a mother. Wasn't heaven beneath my feet? The prophet Muhammad had said, "Heaven lies at the feet of your mother." How could heaven be beneath my feet if I was worthless?

The student continued. He had a friend who had sex with a woman. When it came time to get married, he ditched her for a virgin. "She didn't deserve his respect," he explained. Just two weeks earlier I had raised a point at a meeting of the men who had taken control of the mosque. Not used to being challenged, especially by a woman, the men stormed out of the meeting. The research assistant professor of engineering who had called my father an idiot earlier in the year screamed at me from across the room: "No one respects you! Just leave the mosque!"

That Friday the student argued that Muslim women don't have the same right to declare divorce that men do because "Allah created woman sensitive and emotional, especially during her menstrual period." It is because of this "sensitive nature" that two Muslim women equal one Muslim man as a witness. He spoke against the attacks against Islam by "the enemies of Islam" who use women's rights as "the most convenient" entry point. "They try to corrupt our Muslim women," he said. He equated

"non-Muslims" with "ignorant Muslims" when they criticize the abrogation of the rights of Muslim women. He blasted women's rights leaders in
Egypt as "advocates of hell." Muslim women should not be "deceived" by
Western practices, he said, including "Western women going out to
work." "Who can have . . . our Muslim women . . . follow the West?" It
was right to "prevent intermixing between men and women." To clinch
his sermon of inspiration, he noted: "Any woman who wears perfume so
men can smell her, it is as if she has advocated adultery."

I thought about the body wash I'd used earlier that day. Danny's wife
Mariane had given it to me as a New Year's gift earlier in the year. Aptly
named considering the outrage that consumed me at that moment—as in
so many moments at the mosque—it was called Total Bitch, "for life's little
annoyances." It had a gentle scent of lily, forest foliage, and fern. "Add ex-
plosive suds to a hot-tempered shower," its package told me every morning.
Did wearing the faint scent of Total Bitch make me guilty of zina? Did it
make me a criminal? What had happened in my religion that I had to even
ask myself these questions? How did the scorn for women's sexuality reach
into my personal space in Morgantown, West Virginia? How had women
become so invisible a force in our communities that men could stand at the
pulpit, speaking for us, judging us, and passing edicts about us?

That night I looked at the latest newsletter distributed by the local
Muslim Students' Association. It listed new births, giving the names of
the newborn babies and their fathers. There was no mention of the mothers. We were truly invisible.

I did some research about the assumptions the student preacher had made
about women's place in Islam. My inquiries led me to *Speaking in God's
Name: Islamic Law, Authority, and Women*, a book by Dr. Khaled Abou El
Fadl, a UCLA professor of law. The professor was born in Egypt and
schooled in Western law in the United States and in Islamic jurisprudence in the Middle East. In his pages I discovered that dubious men attributed to the prophet misogynistic statements that puritanical Muslim
men of today use to judge, control, and demonize women.

That Sunday night I went to a meeting of mosque leadership to express
my outrage. My mother met me there after rushing back from my
nephew's soccer game so that she could be there on time. She wore one of
my niece's jackets with letters spelling "Brooklyn" across the chest and
the hood pulled over her head. I was relieved to see her. We were a team.
I wore my sweat jacket with "BrothaHud" across the chest.

"You came!" I exclaimed.

"Of course," she said.

We joined the meeting in the community room. The men were talking about covering the half-inch cracks between the bathroom doors to protect their *awrah,* or forbidden zones, from public viewing. They had already barricaded the windows of the community room with bookcases and a folding table propped up on its side. We sat at a table separated from the horseshoe table where the men sat. I felt like Amina Lawal in front of her jury of men. But there was a big difference. I wasn't alone. And my mother and I weren't there to be judged but to challenge the indictment of any woman as worthless.

My mother lit into the student preacher. "You harassed us for thirty minutes!" my mother exclaimed, her voice even but her outrage obvious. "There was not one positive thing you had to say."

The student protested that he had talked about the rights of inheritance and economic security that Islam gives women. My mother responded: "You lost me at 'worthless.'" She protested his association of adultery with women who wear perfume. I concurred. At that moment the student preacher laid down his cards: "I ask Sister Asra specifically since she raised the question: Does she believe this hadith? If you believe this hadith, you are a Muslim. If you do not believe this hadith, you are not a Muslim."

I laughed. The student had revealed his true colors: relying on the ideology of judgment and the most rigid expression of Islam, he believed he could judge whether a person is a Muslim based on their acceptance of a single hadith. "Your question is unacceptable," I said. "It does not allow for *ijtihad*" (critical thinking).

I had checked the hadith. At the time of the prophet, it was also claimed that the prophet said: "Any woman who puts on perfume, let her not attend the *isha* [nighttime] prayers with us." It was claimed that the prophet said: "Any woman who applies perfume and then goes out among the people so that they could smell her fragrance is a *zaaniyah*" (someone who engages in zina).

I sent an inquiry to the scholars I'd come to know. Amina Wadud at Virginia Commonwealth University wrote back. She noted that the student's claims didn't come from the two highest—and soundest—hadith collections, known as Bukhari and Muslim. "The use of various *hadith* [to discourage] women is a carefully orchestrated methodology," she wrote, whose purpose is to marginalize and belittle women. What about the idea that a woman is "worthless" if she loses her chastity? I got a response from a sexuality scholar, Kecia Ali, a convert who was doing groundbreaking work on women, sexuality, and Islam with Harvard University and Brandeis University. She said, "Chastity and modesty are not the sum total of a woman's worth. That's simply not tenable from the perspective of Qur'an, Sunnah, or jurisprudence."

Dr. Abou El Fadl discredits men like Abu Hurayrah, whose records of the prophet's sayings were often some of the most virulently anti-women elements of the religion. Abu Hurayrah fills pages of footnotes in Dr. Abou El Fadl's book. He converted to Islam late, only three years before the prophet's death. Although he spent less time with the prophet than Abu Bakr (the first caliph), Umar (another caliph), Ali (the prophet's son-in-law), and Aisha (his favorite wife), Abu Hurayrah transmitted more statements by the prophet than any other companion. For that reason, the authenticity of his transmissions has been the subject of debate for centuries, including at the time of Aisha, Umar, and Ali, all of whom severely criticized Abu Hurayrah's reports. Aisha once said to Abu Hurayrah, "Abu Hurayrah! What are these reports from the prophet that we keep hearing that you transmit to the people! Tell me, did you hear anything other than what we heard; did you see anything other than what we observed?" Ever patronizing in making his defense, Abu Hurayrah responded, "Oh mother, you were busy with your kohl [eye liner] and with beautifying yourself for the prophet, but I—nothing kept me away from him."

In the name of the prophet, Abu Hurayrah objectified, marginalized, and hypersexualized women. He said that the prophet declared women were made from a crooked rib, making us more deficient. He was the one who said the prophet declared we would be the majority of the inhabitants of hell in part because our menstrual cycles made us more deficient—and thus worth only half the witness of a man.

Statements of the prophet attributed to Abu Hurayrah had been repeated at my mosque to corral women. I just didn't realize it until my e-ijtihad connected the dots for me. One night the student preacher who had declared that unchaste women are worthless told the men in a study session: "The Messenger of Allah cursed the man who wears women's clothes, and the woman who wears men's clothes." It was another way to plant unyielding gender lines in society—dangerous because they were used to keep women out of schools and the workplace. I looked down at my blue jeans, over which I had pulled an oversized hoodie. "Great, so I'm cursed," I whispered to myself. The source: Abu Hurayrah. It didn't even sound like the prophet to *curse* anyone.

Just about every Friday I heard the prayer leader attribute another statement to the prophet: "The best of the rows for men are the front rows; the worst of the rows for men are the last rows. As for women, the best rows [in prayer] are the last rows, and the worst rows are the front rows." The logic, again, was that a woman's sexuality disturbs a man. And the source of this saying? Abu Hurayrah. Classical jurists, on the other hand, interpreted the prophet's words to mean that there should be a reasonable distance between the last row of men and the first row of women. In our mosque my mother and I sat at the back of our main hall, getting there early, when the hall was mostly empty. Normally we were at least twenty feet from the closest men, in front of us. At the busy Friday prayer we sat in our usual spot for the sermon. Men trickled inside afterward and often took seats in a space behind us. One of the community leaders had told me and my mother that we should pray with our backs against the wall to make ourselves the *last* row. My mother turned to the man and asked, deadpan: "Why don't you just dig a hole, shove us in it, and stone us to death?"

From Atlanta, Georgia, to Cairo, Egypt, women are told that the prophet said, "If a man calls his wife to sleep with him and she does not respond, causing him to be angry with her, angels will curse her until the morning." A Saudi publishing house quoted the saying in a book that's distributed worldwide, *Islamic Perspective on Sex*. My friend Saleemah heard an imam repeat those words to a bride at a wedding ceremony not long ago at the Masjid Taqwa wa Jihad ("mosque of God consciousness and struggle") in the Bronx. "It was clear: she doesn't have a right to say no," said Saleemah, as she related the anecdote. The source of the quote: Abu Hurayrah.

It seems that Abu Hurayrah is to Muslim women's status in Islam what some Christians say the Apostle Paul is to Christian women's rights. In all religions we have a Muhammad or a Jesus who envisioned a world of rights and social justice for women. And every religion has an Abu Hurayrah. One Friday the student preacher admonished the men sitting behind my mother and me to move in front of us, even during the sermon, but he tripped on his words. "The best of the rows for the men are the last; the worst of the rows is the front. As for women, the best of rows for the women are the front. The worst of the rows are the back." My mother looked at me in delight. "Let's go to the front!" But of course we didn't dare. We still couldn't dare claim our space in the front.

At home, shaking my head about the absurdity of the puritans' thinking, I gazed at a recent purchase I'd made on a Muslim website. It was "the

Muslim Doll," a woman named "Razanne." The packaging said, "Ra/zaan def. 1. modest 2. shy." She was the closest I'd get to the student preacher's ideal of the Muslim girl next door. Then I looked closely at the accessories that came with Razanne. I smiled. Sure enough, the evidence against her sat with a hairbrush and a comb; she had a silver-colored perfume atomizer.

I took to the pen again and wrote out my thoughts in an essay that the *New York Times* published on May 6, 2004, on its op-ed page. "Hate at the Local Mosque," the headline said. I called upon moderate Muslims to speak out against the intolerant sermons preached in mosques in America. I exposed the hate sermon and the one in which the graduate student declared unchaste women to be "worthless." My friend Mohja wrote me immediately to support me and lament the lack of involvement by the Muslim world's ACLU. She remembered the verse from the Qur'an that enjoins Muslims to witness against your own people when they do wrong. She also sent me another saying of the prophet: "Help your brother whether he is oppressed or oppressing." When the people asked, "We can help him when he is oppressed, but how can we help him when he is oppressing?" it is said that the prophet replied, "Help him to stop oppressing." Mohja wrote, "I applaud you for doing this. I am dismayed that CAIR is not helping. I think they need to look beyond the lettering of their mandate and see to the heart of this—it weakens their position not to look at things like this. We're supposed to get other communities to help clear the climate of hate when it bothers Muslims, but we are not supposed to campaign against hate when it is spread by Muslims?"

The men at the mosque retaliated in an article that the *Pittsburgh Post-Gazette* wrote about the controversy. One of my chief detractors was a Palestinian American man who accused me of being a publicity seeker. The Egyptian graduate student said that he knew I would take his words about unchaste women personally, but he didn't think I had been appropriately repentant after having committed this major sin. Their comments I expected. What hurt me most was the dig by a woman leader at the Islamic Center of Pittsburgh, an inclusive mosque that had fought off the puritanical marginalization of women. She said that I was acting as a secular feminist, not a Muslim. "To be Muslim is to be a feminist," I told my mother, shaking my head. "And it is America's secular tradition that allows Islam to thrive in America. Just see how many churches exist in the nonsecular Saudi Arabia. None. Pluralism thrives in secularism."

The *New York Times* headline frightened even me, especially because I knew it reflected the truth. I knew we had to be brutally honest in forging tolerant communities if we expected tolerance. And I knew we had to keep asking questions about the source of the hate if we were going to dismantle it. I had no idea my inquiry would lead me straight back to Mecca.

PART SEVEN

HARVESTING THE FRUITS OF THE PILGRIMAGE

June 2004 to October 2004

MUSLIM WOMEN'S RESPONSE

MORGANTOWN—One month later, on the first Friday of June, I realized—albeit for just a day—the kind of sisterhood and brotherhood that I had dreamed about so many times as I stood alone in my mosque. Five Muslim women and one Muslim man descended on my hometown to march through the front doors and into the main hall with me and my father, mother, brother, and son to affirm the rights of women in mosques everywhere.

They were all new friends I had made in the six months since I had decided to take a public stand to claim my rights at my mosque, and the women were all people with whom I could connect. The first one in the group was Saleemah, who had recently left her job at *Azizah* to edit an anthology of young American Muslim women's voices, *Living Islam Out Loud*. Mohja Kahf, the poet, flew in from Fayetteville, Arkansas. Samina Ali, a novelist born in India and living in San Francisco, flew to Washington, D.C., and then drove to Morgantown with Sarah Eltantawi, an activist and writer living in New York.

Born in California, Sarah had risen to become the public affairs director at a national Muslim organization, the Muslim Public Affairs Council. She battled the neoconservatives on talk TV from CNN to Fox News. And her bright red lipstick always seemed to be freshly applied. Her inspiration was her mother, Hoda Eltantawi, an assistant hospital administrator in southern California and former business owner. For Sarah, her mother was the most insightful, dignified, and elegant person she had ever known, a woman who seemed to be totally self-sacrificing while never losing sight of herself.

At the last minute, we had recognized that we were nothing without our mothers. Saleemah invited her mother, Nabeelah Abdul-Ghafur, a community activist and writer from their hometown in New Jersey. And I insisted that my mother join us. Besides our march, we planned an evening literary event sponsored by the Morgantown Public Library—a fitting sponsor, since the library had played a critical role in my independence and empowerment. I remembered as clearly as yesterday scouring its shelves in the spring of my freshman year at West Virginia University to find a magazine where I wanted to spend the summer interning and coming across *Harper's*. The Center for Women's Studies at West Virginia University was also a sponsor, and that meant a lot to me. The son of its founder, Judith Stitzel, was the boy with whom I had square-danced in sixth grade. Another sponsor was meaningful to me: the Shelley A. Marshall Foundation. Shelley was the wife of a classmate of mine from Suncrest Junior High and Morgantown High, Donn Marshall. She had died when terrorists flew a plane into the Pentagon on September 11, 2001. "Shelley wouldn't have ever accepted a back door," her husband said. "She would have supported you."

Before the group's arrival, I worked into the night to create a brochure for our event. I struggled with a name until I finally settled on the perfect expression of what we were doing: "The Daughters of Hajar: American Muslim Women Speak."

The day before our march, the Associated Press's Allison Barker sent out a dispatch in which my mosque said publicly for the first time that women could walk through the front door and pray in the main hall. The mosque leader: a woman who had become the mosque spokeswoman. We had had a small victory at the mosque. In a May election, the first woman ever was elected to office to something we called the executive committee. A young American convert, she was a lawyer, and we had bonded over our mutual hopes for an inclusive community in which we could raise our children. We were running together for office. In a dramatic moment at the mosque, the conservative men included her on their slate for election. They voted against my being on the slate. I wouldn't have accepted a position on their slate. I turned to the lawyer: "Are you sure you want to be associated with them?" She wanted to try to bring about change from within. I didn't begrudge her choice even though I felt betrayed. That night, she prayed against the back wall where the most conservative men wanted my mother and me to stay. My mother and I prayed in front of her. The men had accomplished what they wanted: division. I withdrew from the race. She won office and so did three moderate men. "It's a jump ball," I told a friend. We had won the struggle to open the doors of the mosque to women. We were Muslim women *redefining* the boundaries established to control and define us.

Standing proudly with us was a young Muslim man who had driven from Albany, New York, to support our effort. His name was Michael Muhammad Knight, and a punk novel he wrote had first presented to me a scene where a woman leads men in prayer. Then I discovered that in the seventh century the prophet had given a woman named Umm Waraqa ("mother of Waraqa") permission to lead her staff in prayer, including a man who happened to be a slave. Most Muslims believe that a woman can't lead a man in prayer, but some scholars, such as UCLA law professor Khaled Abou El Fadl, argue that the Qur'an establishes no moral hierarchy between men based on whether they are slave or free. Thus, the prophet's designation of Umm Waraqa as a prayer leader over a congregation that included a man established two points: that the determination of who should lead prayers should be based on religious knowledge, and that a woman could be designated to lead prayers in a group that includes men.

The other boundary around us was the notion that women must pray separate from—or at least behind—men. Of course, as I had seen, this wasn't practiced in the holiest of Muslim cities, Mecca, and religious scholars such as Dr. Abou El Fadl note that at the time of the prophet men would assemble in the front of the mosque but late-arriving men would fall into prayer lines behind and beside women. A little-discussed but important piece of historical context also makes the issue of proximity between men and women more complicated: men and women didn't wear underwear or trousers in those days, and proximity easily led to immodesty. Modern hijab, adopted for reasons of modesty, makes such considerations irrelevant. As evidenced in Mecca, today men and women can pray shoulder to shoulder and even with women in front of men with no fear of indecency.

Assembling appropriately at the West Virginia University School of Law, as we sought justice, we planned to redefine the boundaries that had disempowered us as women of faith in our communities. If I had to wonder about our correct path, my nineteen-month-old son, Shibli, was reassuring as he lay in full prostration, unprompted, while we sat in prayer, led by Saleemah's mother, Nabeelah. I felt at that moment that we had divine blessing for the action we had just taken. We were eight women spanning two generations from New York City to San Francisco—writers, mothers, sisters, and poets, born in India, Syria, California, and New Jersey, converging in this university town in the Appalachians. We were the physical manifestation of a reform quietly occurring within Muslim communities in America. A taboo act in most communities, Mike

Muhammad Knight stood in line with us, my mother to his right, and he followed Nabeelah in prayer.

We stood and walked down Law Center Drive, crossing the street onto University Avenue, passing Sanders Floor Covering, Hartsell's Exxon, and the golden arches of McDonald's, chanting the call to prayer that I had declared the year before on the holy pilgrimage of the hajj. "Here I come," Saleemah said, leading us.

"Here I come!" we responded.

"At your service, oh Lord," Saleemah continued.

"At your service, oh Lord," we repeated.

To proclaim a united purpose with other Muslim women was important to me because I had very much lacked that kind of connectedness in the world. When Saleemah paused during our march, Mohja started us in a second prayer said by the prophet when he entered a mosque: "God grant before me light, and behind me light, and on my right light, and on my left light, and above me light, and beneath me light, and grant me light."

Shibli rustled in my arms, light upon me. I no longer wondered whether my back was heavy from the burden of the sin that I had committed in creating my son, as I had thought in Mecca. Our voices rose in an exaltation that surprised even me as we marched to the front door of my mosque and posted a symbolic message on its door by walking over its threshold and ascending into the main hall: women have an Islamic right to equity in the Muslim world, and we will no longer accept the marginalization imposed by cultural traditions. Inside, we stood together, and I felt the press of many shoulders beside mine. In the same line but just a little bit apart, to placate sensibilities, Mike stood beside us. As prayer began a remarkable thing happened: a teenage boy stepped into our line beside Mike. There was hope for a new future.

FROM THE MOSQUE TO THE BEDROOM

MORGANTOWN—As the issue of my sexuality repeatedly became a lightning rod for criticism of my effort to reclaim my right as a Muslim woman to public space, it became clear to me that we have to also reclaim our rights in the most private space in our lives, the bedroom. In my case, that means reclaiming my Islamic right to be free from punishment for having had the premarital sex in which Shibli was conceived, even if it's considered morally wrong; my right to keep my baby; and my right to be free from gossip, slander, and judgment about decisions I've made about my body as an adult. From public space to private space, followers of the puritanical brand of Islam have tried to control women. We have to reclaim our rights from the mosque to the bedroom.

Sexual authoritarianism in the name of religion has been practiced by the orthodox followers of all the religions of the world. It most often expresses itself as controls over women's bodies. The Catholic Church has declared that the sole purpose of sex is procreation and made abortion grounds for excommunication. Orthodox Judaism requires the evidence of blood on the bed sheets from the bride's ruptured hymen after her wedding night. Puritanical Hinduism doesn't allow widows to remarry, even if they are widowed at a young age, thus denying them sexual intimacy for the rest of their lives.

What I realized was that sex has often been used to deny women not only their sexual rights but their religious rights, such as at my mosque when some of the brothers claimed that women would cause fitnah, or conflict. Sexual and religious rights are also intertwined in the issue of second wives. Sisters in Islam, the Malaysia women's rights group, lays out the rights of Muslim women with regard to polygamy. Islam didn't invent polygamy, it notes, a practice common in pre-Islamic Arabia, and the Qur'an, in fact, was restrictive, not permissive. "If you fear that you shall not be able to deal justly with the orphans, marry women of your choice, two, or three or four; But if you fear that ye shall not be able to deal justly (with them), then only one. . . . That will be more suitable, to prevent you from doing injustice," says the Qur'an in "Al-Nisa," (The Women, 4:3). Elsewhere, the same chapter (4:129) notes: "You are never able to be fair and just as between women, Even if it is your ardent desire . . ." A Muslim woman has the right to protest a husband's desire to marry a second wife, but even in America Muslim men in places like Morgantown sometimes try to keep their first wives compliant by threatening to take a second wife. One afternoon a Muslim woman whispered to me, "*Please* write about second wives." "What about them?" "How they are *not* allowed." Not even in Morgantown.

When Saleemah, Samina, and I wrote the vision statement for the Daughters of Hajar, we addressed sexuality at the same time that we campaigned for access to mosques. Our priority was to publicize the rights that Islam grants to women regarding their sexuality and to tackle sexual taboos, such as homosexuality. Mohja imagined, for instance, a "wedding night" initiative to better prepare women for what is often a first-time experience with sexual intimacy.

Reclaiming women's rights from the bedroom to the house of worship isn't a challenge unique to Muslim communities. In 1963 Betty Friedan helped usher in the second wave of the American feminist movement with *The Feminine Mystique,* a book advocating equal rights for women from the bedroom to the world beyond. Reclaiming Muslim women's sexual rights isn't a new effort either. It dates right back to the seventh century and the start of the Islamic faith. As I see it, the Qur'an rejects the sexual double standard so often imposed on women by giving Adam *and* Eve equal responsibility for their exile from heaven. It celebrates sex by declaring poetically that men and women are garments to each other. It speaks sensually about a man's seduction of a woman being like the nurturing of a farm field.

But in most corners of the Muslim world we have long departed from respecting women's equal rights in the bedroom. Divine law has often been used to sanction male promiscuity. From Florida to our campus of West Virginia University, young male Muslim foreign students have sex with women, often American, in something called temporary marriages. They were allowed by the prophet in times of war, but it's a stretch of the imagination to apply that criterion to college campuses in the twenty-first century.

This code has many rules that are accepted by puritanical Muslims, who use a woman's sexuality to sentence her to a prison of silence, shame, and subjugation. The first is that we should be strictly separated and segregated. The second is that we should remain silent and submissive.

There is a common denominator in this debate about women: fear. One day outside my house a chemistry professor from Bangladesh protested my actions at the mosque. "I don't want AIDS," he told me.

"Excuse me?" How were women's rights in the mosque related to AIDS?

"It will lead to unnecessary mingling."

"You think we're going to get AIDS sharing the same space in a mosque? We're not going to be jumping on each other just because we're in the same room," I said. I quoted the Fiqh Council of North America's ruling that men and women could mix in the mosque because it's the best place for us to learn how to interact Islamically in the outside world. He didn't care.

"It will lead to unnecessary issues. Look at American society. Look at the way men and women live together having sex without being married."

My father lost it. "And the Arab sheikhs that you follow? They're any better? They buy girls in India and Lebanon and God knows where else, have sex with them, and throw them away!"

Then the chemistry professor's teenage daughter bounced down the hill with a tennis racquet over her shoulder.

I wanted to do a Kinsey report on sexuality in the Muslim world. Dr. Alfred Kinsey broke new ground in 1948 when he and his colleagues at Indiana University's Institute of Sex Research (now the Kinsey Institute) published a controversial and groundbreaking study, *Sexual Behavior in the Human Male*. The report found that 50 percent of married men had affairs; that as many as 10 percent of American men were predominantly homosexual; and that 37 percent had engaged in a homosexual act of some sort during their lifetime. Some American religious conservatives accused Kinsey of being part of a Communist plot to undermine the American family, but a Gallup poll found that 78 percent of the people surveyed considered Kinsey's research "a good thing." The media compared the report to the explosion of the first atomic bomb three years earlier, and America held its breath as it waited for Kinsey's report on women's sex lives. When that report appeared, Kinsey inflamed conservative critics when he said that "it is the church, the school, the home, which are the chief sources of sexual inhibitions, the distaste for all aspects of sex, and the feelings of guilt which many females carry with them into marriage."

The dichotomy of the private world versus the public world in Muslim communities, as in many traditional communities, leads all of us to avoid being completely honest about our sex lives as Muslims. For all of the judgment against Muslim women who have premarital sex, how many Muslim men do as well? For all of the judgment against Muslim women as sexual beings, how many Muslim men have affairs or use polygamy, temporary marriages, and other forms of religious cover to get extra action in the bedroom?

Kinsey's findings challenged many of the essential assumptions about sex and gender that defined American society at the time. It blew open the lid on many myths. I imagine that a report on the sex lives of Muslims would very much do the same for Muslim society. We could learn from the issues that emerged in the West with the empowerment of women— such as women's tendency to burn out when they strive to be superwomen who have it all—or we could also just go through our own growing pains as we mature as a Muslim society.

INQUISITION

"There goes a woman," resumed Roger Chillingworth, after a pause,
"who, be her demerits what they may, hath none of that mystery of hidden
sinfulness which you deem so grievous to be borne. Is Hester Prynne the
less miserable, think you, for that scarlet letter on her breast?"
. . . The scarlet letter was her passport into regions where other women
dared not tread. Shame. Despair. Solitude! These had been her teachers—
stern and wild,—and they had made her strong.

Nathaniel Hawthorne,
The Scarlet Letter (1850)

MORGANTOWN—During a lifetime as a reader, I had never taken to Western literature, but all of a sudden, in the summer of 2004, I was absorbed in every detail of American author Nathaniel Hawthorne's novel *The Scarlet Letter*. It tells the story of a woman, Hester Prynne, who is sentenced in Puritan America to wear the letter "A" on her bosom for the act of adultery that created her daughter, Pearl. I felt as though Hester's story was my story. The men at the mosque were putting me on trial.

In my case, events unfolded like a twenty-first-century e-inquisition. In the quiet of a Tuesday afternoon I received an e-mail with an innocuous subject line that included a typo: a memo concerning "a Petition and a meeting of the ExecutiveCommittee of the ICM." The president of the mosque executive committee, the engineering professor who had led the takeover of the mosque earlier in the year, notified me that "we have received a petition, on Friday June 18 2004, signed by about 35 members of the ICM asking to revoke your membership and expel you from the ICM."

The petition cites several charges that alleges that you have engaged in actions and practices that are disruptive to prayer, worship and attendance at the ICM, and that you have engaged in actions and practices that are harmful to the members of our community.

I picked up the phone and called my mother. "They're doing it. They're trying to kick me out of the mosque."

The engineering professor lived five blocks away from my parents' house. His wife had invited me over for tea to tell me to stop my actions of protest at the mosque. Naively, I had thought she wanted to privately convey her support for my fight for women's rights. Instead, she admonished me. "The men are selfish. Let them have their space." She told me that I should spend the rest of life seeking repentance for having had my son while I was unmarried.

Later the professor had complained about "the sound of Ms. Nomani writing notes and flipping pieces of paper" during Friday sermons. In a twist of fate, the professor had been elected president of the executive committee after an election rushed through the mosque. The students who had preached intolerant sermons at the pulpit were now imam at the mosque. And I was on trial. The engineering professor gave me two days' notice to appear before a 9:15 P.M. drawing for the names of people for my mosque tribunal.

I asked for a copy of the complaint against me. Denied. I asked for the names of the complainants. Denied. I asked for a delay so I could find legal counsel. Denied. The Supreme Court had that week granted new rights to Muslim prisoners in Guantanamo Bay, but I was being denied the same elements of due process that Muslim civil rights groups were rightly demanding for Muslims.

My friends rushed to support me. Mohja dispatched an e-mail to the engineering professor and tried to appeal to reason—and Islamic precedent.

What happens to Muslim women of achievement has been recorded by sociologist Dr. Aminah McCloud: We are pushed away from the Muslim community. Used to holding respected position and influence in the secular world, Muslim women who are successful professionals are unable to accept the banishment to the rear entrance and the marginal influence that is usually the lot of women in our mosques, and that more traditional Muslim women do not find objectionable.

Thus the Muslim community loses women who could, instead, be among its most ardent contributors and leaders.

I am writing to urge you to dismiss the petition to ban Ms. Asra Nomani from the Islamic Center of Morgantown.

She did not take her issue public until she had corresponded with the mosque board president at the time. . . .

In banning her, you will be losing a potential resource.

Furthermore, the activities spoken of as the cause of the ban are over. They are in the past. The struggle is finished; why not let it rest? Why this malicious last effort to "win," when clearly you (the mosque leadership) were wrong (about not allowing women in the main hall

or through the front door) and you have corrected your wrong, after
long resistance? It is as if you are ashamed to admit that she has corrected you in a wrong, the way a woman corrected Caliph Umar. But
his reaction was simply to say, "A woman was right and Umar was
wrong."

Do you and your board brothers have the moral character to admit
you were wrong and let it go like Umar *radia allahu anhu* [may Allah
be pleased with him]? Do you have the courage to refrain from allowing this malicious action (the ban) from becoming a poor ending to
the story of your story with Ms. Nomani? Can you help your brothers
on the board clean their hearts of rancor and bitterness? Do you have
the largeness of spirit to draw Ms. Nomani into your community as its
friend, a woman brave enough to stand for justice even if it made her
unpopular, and to see her now as a potential ally, in a new morning?

These are the questions you face. I pray you and your community
will be granted wisdom by the Merciful One to answer them well.

The answer came back swiftly, full of typical e-mail typos. The engineering professor wasn't going to budge. And he was again arguing that I was
to blame for speaking out.

we believed that good muslims should always resolve their problems between themselves without resortinto the media because in this
case we loose much more than we can ever gain with the known bi-
ases in the media agaist Islam and muslims,

as for the petition, I must say to you that I am obliged to follow the
constitution and start the process which will take its due course, . . .

May Allah SWT ["the sacred and the mighty"] guide us all to his
striaght path and let our work for his sake only,

There were two words that separated my fate from Hester Prynne's: *pro
bono*. "Time to hire a lawyer," I told a friend.

"Wow," he responded. "Did Jeanne d'Arc have a lawyer?"

She didn't. In the fifteenth century, Jeanne d'Arc, a village maiden, inspired and led the soldiers of French king Charles VII to defeat British
forces who were keeping him from power in the Hundred Year War. Captured, she was handed over to French clerics of the Catholic Church who
supported the British and put on trial for witchcraft and heresy. During
her trial sixty highly trained politicians, lawyers, and ambassadors grilled
nineteen-year-old Jeanne, even subjecting her to examinations by women

to check whether she was a virgin (she was). Alone, she defended herself, refusing to answer questions she was asked twice and telling court reporters to record her previous answers. During fourteen months of interrogation she was accused of wrongdoing for wearing masculine clothes and of heresy for believing she was directly responsible to God rather than to the Roman Catholic Church. She was burned at the stake on May 30, 1431. Pope Calixtus annulled the verdict in 1455, calling it wicked and unjust. Jeanne d'Arc was canonized in 1920; her life is now celebrated with a feast every May 30, my niece's birthday.

Like Jeanne d'Arc, countless other women throughout history have been unfairly persecuted for challenging the status quo. I knew one thing: I was going to fight, but I wasn't going to be a martyr. I was going to find a lawyer.

I wrote to David Remes, an attorney whom I had found after returning from Pakistan to help Mariane, my friend Danny Pearl's widow, and asked him if he knew a good First Amendment attorney. To deny me access to my place of worship was to deny my right to worship with the only assembly we had in Morgantown. And to punish me with exile because I spoke up about injustices and intolerance was an attempt to squash my right to free speech. David was a partner at a high-powered Washington lobbying and law firm, Covington & Burling, and he was one of the defense lawyers for Muslim prisoners being held at Guantanamo Bay. To my shock, the next day he said that he would represent me. "You will?" I asked, doubting that I had heard right. Having absorbed so many of my community's messages of illegitimacy, I didn't feel I was worthy of help. "I will," he responded, clearly. David, a Jewish American born in New York City, told me later, "What moved me to help you was my general intense sympathy for those who are powerless and alone and who are being punished for simply being themselves. The greatest cruelty is punishing people for being who they are. Conversely, I intensely dislike the idea of 'correct' thinking. Freedom of mind and expression are my creed." For the first time I didn't feel alone. And I realized the incredible power of legal muscle when you have it on your side.

My lawyers couldn't arrive in time for the scheduled first proceeding, and the elected mosque president refused to delay it. On the night of the meeting a searing pain ripped through my head, and I felt nauseated. In the minutes before my trial was to begin, I wanted to flee. My cell phone rang. It was Amina Wadud, the African American scholar of Islam who had taught me the Islamic concept of *tawhid*, the oneness of God with all things created—equality. "They have no Islamic right to banish you from the mosque," she told me firmly. "Stay strong." I lay down at home, trying to draw on whatever reserves I could find. Instead, I found myself doubled over in pain on my bed. "They are trying to drive me from my home," I cried. "I feel so alone." But I wasn't. As the moment arrived for me to leave, my father embraced me and said, "You are not alone. We are here." My mother too had stood by me on my journey. "This does not upset me," she said, looking me straight in the eye. "It should not upset you." With that, she gave me the simple jump-start kick only a mother can give a child. I splashed cold water on my face and knew that I would not run away.

Sadly, as I went to the mosque with my mother and Shibli, my father, a man who had helped to build the first mosque in Morgantown, was on his way across town to Ruby Memorial Hospital. The stress had caused him to literally lose his breath, and his doctor wanted to check him out. As I later learned, he slept with electrodes attached to his body while his brothers in the mosque were putting his daughter on trial.

I walked up to the front doors of the mosque, took a deep breath, and stepped inside. I took a seat at one end of the cafeteria-style tables arranged in a horseshoe shape in the community room. An assembly of my community sat around the table, mostly men with beards, crocheted prayer caps, and dimly colored pants and T-shirts; others were clean-shaven, intermixed with women hooded with hijab. I tucked my jet black hair into the hood of an oversized, black hooded jacket I had won in a beach volleyball tournament in my younger days. Like Hester most of her life, hiding her lush hair under a cap, I was making myself asexual in this world in which my sexuality had become the evidence of my criminality. My jacket carried the label "Six Pack"—insider volleyball lingo for the power of a hard-driven spike hitting an opponent's face.

The professor pulled strips of paper with names on them out of a plastic bag. He read the names as if he were calling the winners of a raffle. In fact, the people named were going to be the jury in the secret mosque tribunal. The judges would be the five-member board of trustees whose president had ordered me away from the front door the October before. Now, months later, while the world watched Saddam Hussein's defiance on the judicial stage, I wasn't allowed to even turn on my digital recorder. When the professor had finished drawing names, he proclaimed, "We're done with this, and we need to go on." Sadly for my detractors, *this* wasn't going to go away that quickly.

Both the world and the Muslim ummah are at a crossroads in history. Within the Muslim world, we must open the doors on *ijtihad*, not slam the doors shut on critical thinking for the sake of political correctness. To try to close the doors of God's house to me through a secret mosque tribunal was to try to close the doors on truth, justice, and tolerance. I had decided to fight this effort to ban me to defend not simply my right to walk into my mosque but the highest principles of Islamic teachings—the truth and critical thinking.

That night the mosque leaders reduced me to lining up the slips of paper in an effort to check the names against our membership list. Only my mother and son were with me. When the elected president, derisively dismissing my mother's insistence that I was entitled to a lawyer, said, "Sister, this is an *internal* community matter," my mother pounced. "Don't call me *sister*. You don't treat me like a sister."

The West Virginia convert who had organized sisters' swimming now was chairwoman of the *dawah* committee, and she tried to talk to me as I tried to put the papers in order. She used to be the bell ringer. Now she had won a position at the table after all of my clamoring, and I had heard not a peep from her. She whispered to me, "Asra, I want you to know that a lot of people don't support this petition to ban you."

That could very well be true, but I did the math. About thirty-five members had signed the petition. That left about one hundred members, including this woman, who hadn't taken a strong stand against them. They were the silent moderate majority in the larger Muslim world who allowed the extremists to define Islam in the world. "Great dawah they're doing for Islam," noted my mother, all too prescient for the liking of our community. This chasm between how Muslims act and what Islam teaches is the kind of challenge that all societies have had to face. As so well described in *The Scarlet Letter*, people distort religion and do it a great disservice when they punish people for their moral judgments. The challenge for each of us is to discern our own personal faith from the doctrines others try to impose on us.

Nina Baym, a professor of English at the University of Illinois at Urbana, said in an introduction to *The Scarlet Letter*: "Although Hester suffers enormously from the shame of her public disgrace and from the isolation of her punishment, in her innermost heart she can never accept the Puritan interpretation of her act. To her the act is inseparable from love. . . . Because she does not believe that she did an evil thing, she retains her self-respect and survives her punishment with dignity, grace, and ever-growing strength of character." These words could have been spoken from my own heart. The conflicts that Professor Baym identifies in Hester Prynne's life are the same ones that defined mine: private versus public life, the spirit versus the letter, the matriarchal versus the patriarchal ideal. Hester overcomes the power of the letter of the law to believe in the spirit of nature. She stands up for the freedom of her inner world over public expectations. And she is a triumph of matriarchy over the constrained patriarchy of the Puritan elders.

One of the issues in my trial was the allegation that I aired our community's dirty laundry and shamed the community. I had tried to get the attention of the mosque leaders, but they dismissed me without even a meeting. The men at the mosque wanted me to absorb their projected messages of illegitimacy. I couldn't, though. To do so would have been to dishonor not just my son but the force that had created him. I knew this effort to ban me would also be a de facto ban on not only my son but the future of truth-telling in the Muslim world. I knew, however, that truth and righteousness would prevail. My son was living proof.

The meeting broke up with the azan, which starts "Allahu Akbar." Shibli had first heard it whispered into his ear by my father. Then he heard it in the holiest places in the Muslim world—Mecca, Medina, and Jerusalem. In the nine months since I started taking him to the mosque, he had heard it called out by the graduate student in engineering who described unchaste women as "worthless." Listening to the call spill into the room, Shibli paused from spinning around the room with his Foosball balls. "Ababooboo," he said through a grin, imitating "Allahu Akbar." Then, as my mother and I prepared to leave, Shibli scampered upstairs into the main hall and stood in the sacred space where my mother and I had taken to sitting. He lay in full prostration in the darkness, then jumped up with a smile and caught my hand in his own.

On a whim, I arranged to talk to Khaled Abou El Fadl, the scholar of Islamic law at UCLA whose books had so helped me understand the misogyny in religious traditions. As I talked to him, Dr. Abou El Fadl quoted from the Qur'an (4:135) and set my spirit free:

Bear witness to the truth even if it is against yourself, your mother, your father, your tribe.

Dr. Abou El Fadl said Islamic jurisprudence does not allow political expediency to override a morally compelled duty such as speaking the truth.

"They concede all of that, but they say because of political considerations, be a partner in a conspiracy of silence. But the prophet says a person who conspires is a silent devil," he told me.

The complaint against me had come like an act of retribution after the leaders of the mosque had finally given women new rights. "No one has a right to put you on trial," said Dr. Abou El Fadl. "It is nothing less than an inquisition."

At home, my father did something extraordinary. He, an elder, testified to the truth. He took pen to paper and wrote an essay for the website Beliefnet, appealing to the Muslim men of our communities to open the doors, not close them, to his daughter and other women. "I am the patriarchy that feminists talk about in women's studies courses. I am the status quo. I am the old guard," he wrote. He explained that the inquisition was unprecedented in his lifetime of working within the Muslim community, but "it is emblematic of the way that extremists and traditionalists try to squash dissent within the Muslim world." He wrote: "Turning to the power of the pen, my daughter was alienated for exposing the community's 'dirty laundry.' But someone had to try to clean it, not continue to stuff it into a corner."

In response to the narrow-mindedness of the mosque leadership, my father had resigned from the board. "I am an old man. I don't have energy to argue and fight with stubborn people," my father said. "Many of the leaders at the mosque want to continue native traditions followed in the U.S., disrespecting the human rights of women. They need to be more open and tolerant not only towards women, but also to those who aren't Muslim and those who don't follow their particular ideology."

After retiring, he wrote that he was happy to be a soccer mom to his grandchildren. "Sadly," he said, "I feel as if I was a failure in protecting women's rights at our Islamic center. Other men, in all communities, remaining locked to tradition and power, need to transcend their egos so they can understand the pain and suffering women endure at the hands of inequity and injustice." He made one request: in the case of his death, he asked that his prayer be performed at a mosque where women are respected. And he made an admission: on the day of our march to the mosque, he walked inside—behind my mother and me.

BACK TO MECCA

MORGANTOWN—As I waited for my trial to proceed, I continued going to the mosque. The last Friday in June 2004, the Egyptian graduate student in engineering stood at the pulpit and shocked me with his gall. He noted: "Allah forbade the believers from loving or caring for the disbelievers and stated that doing so leads a Muslim to become one of them which Allah forbid." I couldn't believe my ears. I immediately thought about my Jewish and Christian friends whom I deeply cared about. I pulled out a paper and started scribbling his every word, not even discreetly this time.

Allah said, he claimed, "Do not take the Jews and the Christians as allies." Oh, that made a lot of sense. What, then, was he doing in America? How could Saudi Arabia have an alliance with America? He continued: "One of the most important fundamentals of our religion is to love and be loyal to Islam and the Muslims and to hate and renounce the disbelievers." He had twisted the first chapter of the Qur'an, adopting a translation commonly used by Wahhabi clerics, to proclaim that Jews and Christians have strayed from "the straight path." This path seemed to be very different from the one my mother had taught me about since my childhood days. I knew many Jews and Christians who lived on "the straight path" more religiously than some Muslims I knew—including this man, who was hurling insults, it seemed, at everyone he could. To punctuate his point, he called Jews and Christians "cursed."

I knew this ideological bully. He was the same student preacher who had called unchaste women "worthless." I sat in my usual spot in the rear right side of the main hall, aghast. Since I had penetrated into the sacred men's space, Friday after Friday I had protested the sermons preaching intolerance against the West, Christians, Jews, Hindus, and women. They were professors, graduate students, and engineers at West Virginia University who stood at the pulpit with hate-filled words rolling off their tongues. I noticed that sometimes they stumbled on their words and didn't even seem to be able to read the lectures printed out on sheets of paper in front of them. Most often, I protested alone. But this time, fortunately, so too did others. The elected committee had an emergency meeting and did the right thing: they fired the student from his post as a voluntary imam.

That weekend he wrote his defense to the congregation and provided a startling revelation: he had gotten the sermon from a website called al-Minbar ("the Pulpit"). At my desk at home I plugged the site's name into my Web browser. What I found stunned me. The site was a slick one that disseminated prefab speeches of hate written by clerics in Saudi Arabia. It called itself the "orator's garden and the Muslim's provision." It followed the Wahhabi school. The speech I had just heard in my mosque was a combination of sermon number 1628 and number 1381. As I scrolled through the sermons, I read verbatim the other sermons that had been delivered at my mosque. In his earlier speech denouncing unchaste women, the graduate student had referred to "a friend" who married a virgin rather than the girlfriend with whom he had had sex because the girlfriend didn't deserve his respect. The student had no such friend. The incident was lifted straight from sermon number 678. The Egyptian American WVU engineering professor who was overseeing my trial had delivered a sermon that spring about the West being on the "dark path." That was sermon number 184. I found the source of the words by the graduate engineering student from Saudi Arabia who exhorted the congregation in April to "hate those who hate" the prophet Muhammad, right down to the mention of a "filthy polytheist": sermon number 848.

Recalling the famous dummy of the vaudeville ventriloquist Edgar Bergen, these Muslim preachers in the United States had become the Charlie McCarthys of Islamic fundamentalism. But who was pulling their strings?

I turned to Internet sleuthing techniques I had learned trying to find Danny's kidnappers in Karachi. E-sleuthing is a new and critical component in the war to stop terrorism as criminals increasingly use the Internet and false e-mail addresses to hide their identities. The one difference here was that al-Minbar proudly named its top officers and even provided a cell phone number. It also displayed its home base prominently: "Direct from Makkah." That was the same Mecca—birthplace of the prophet, holiest city of Islam, heart of the religion—where I had walked only the year before in my spiritual pilgrimage. I was stunned.

The Saudi cell phone number didn't work, and my messages sent to the e-mail addresses listed as contacts went unanswered. I called the Saudi embassy for comment and sought out the spokesman, Nail al-Jubeir, who had been an unexpected ally on the issue of women's rights in mosques. He remembered me. He listened to my findings about al-Minbar and immediately expressed outrage at the messages in the sermons, but he was no dummy. His brother was Adel al-Jubeir, Crown Prince Abdullah's foreign affairs adviser and a top deputy to the Saudi ambassador to the United States, Prince Bandar bin Sultan, and they were both on the TV circuit regularly touting the Saudis' alliance with America in its war against

terror. In the remarketing of Saudi Arabia after 9/11, a new thirty-second
TV ad, "The People of Saudi Arabia: Allies Against Terrorism," showed
images of women working in labs and in front of computers and conveyed
a simple message: "In thirty years, Saudi Arabia has changed from a desert
kingdom into a modern nation. Our nation's citizens are educated and
committed to a more secure future. We are proud of how far we have
come in thirty years and look forward to strengthening our alliances." It
ended with an image of two girls sitting on a rocky ledge by the water,
their hair flowing freely in the wind—an experience that my niece, let
alone my mother and me, would never have been allowed to enjoy in
Saudi Arabia because of the contradictory reality of that rigid culture. I
noted the condemnation in the al-Minbar sermon of Muslims being allies
with the West—an injunction clearly contradicted by Saudi Arabia's
friendship with the United States.

The press aide immediately tried to distance himself from the website.
"Just because they say they're in Mecca doesn't mean they're in Mecca."
He said he would have his technical staff look into it. His assistant wrote
back that the website's Internet server was in Virginia. That was a mean-
ingless fact. A server's location has no direct association with the location
of the *providers* of content to a site.

Indeed, the Saudi government was on the wrong trail. I logged onto a
website database company, Network Solutions, and found out that, according to Internet database records, al-Minbar was run out of Mecca,
Saudi Arabia, and was associated with the Internet address 207.13.11.2.
Another database revealed that the IP address belonged to an innocent-
sounding company called Infocom Corp. It ran al-Minbar's website on a
server in Richardson, Texas. I got no answer to my inquiry at Infocom; an
e-mail to the company bounced back. To my surprise, I found a mention
of Infocom on the U.S. Attorney General's website. It turned out that the
company's founders, the Elashi brothers, had just been found guilty of numerous federal charges brought by the U.S. Attorney's Office in Texas related to money laundering and aiding and abetting a terrorist.

The trail to al-Minbar was getting murkier and murkier.

Somehow, from Morgantown, I returned electronically to Mecca. An
e-mail to al-Minbar came back with an answer from a man named Sheikh
Wajdi al-Ghazzawi, a father and entrepreneur. He was listed as the executive director of al-Minbar. He readily agreed to a phone interview, during
which he connected the dots on a communications network tied to the
Saudi government. He ran his own website with photos of himself at
work. At the age of thirty-nine, he was exactly my age. He was an ambi-

tious man who spoke impeccable English. He was born and bred in Mecca and in fact worked on the outskirts of town in a neighborhood I had passed on my way into Mecca for my pilgrimage. It wasn't far from a storefront I had passed with a misspelled sign, the Center for Islamic Propagation, which was one of the major dispensers of the Wahhabi propaganda that filtered into the world from Saudi Arabia.

Sheikh al-Ghazzawi was also the executive director of a company called Webcrescent. He listed as his clients the Saudi-run Islamic University of Medinah, the Saudi Ministry of Religious Affairs, the Office of the Amir of Jeddah, and, close to home regarding the work that I was doing, the Saudi Ministry of Hajj. I had crossed paths with all of these entities during my hajj. Sheikh Alshareef, our group guide, had graduated from the Islamic University of Medinah. The other government offices had coordinated our hajj, as they did every hajj every year. Sheikh al-Ghazzawi started al-Minbar in 1999 as a way of spreading the faith. He was part of a growing rank of Muslim ideologues who had earned the label "cyber mujahideen," a brand of freedom fighters waging their ideological battle in cyberspace. While the United States and other nations vigilantly watches its ports, airspace, and water supplies, it faces more insidious opponents in cyberspace, where lethal ideas cross borders without censor. Some of the treasures from al-Minbar sermons, all originally preached from the pulpits of Saudi Arabia, include:

- Christians and Jews are "enemies of Allah."

- Jews are descendants of "apes and pigs."

- "Muslims must educate their children to *jihad* . . . and to hatred of Jews and Christians."

- "Muslim women's rights are a Western ploy to destroy Islam."

Ironically, Sheikh al-Ghazzawi was particularly proud of one of his latest creations: an antiterrorism website called TerrorExposed.net for a Saudi government office. The site quoted fatwa by the Wahhabi cleric who had written my pocket-sized prayer book and the other book I had found denouncing nonbelievers as people who should be killed. When I was checking out the site, his politically correct fatwa was that Islam doesn't allow the killing of innocent civilians. I considered this to be true, but I still thought it was most unfortunate that clerics like him were allowed to distribute their work spewing intolerance and hatred. Interestingly, Sheikh al-Ghazzawi said that al-Qaeda operatives considered him an enemy because he posts sermons that denounce the killing of civilians. Al-Minbar abided by a strict code for sermons: they could not be political; they could not name enemies; and they had to be grounded in the Qur'an and the Sunnah.

How, then, could he justify the rhetoric of hate found so frequently on his website against Christians, Jews, and others—including Muslims—who didn't follow a specific interpretation of Islam? The Qur'an, he insisted, condemns acts but not people. "Like hating the sin but not the sinner?" I asked. "Exactly."

So I read from the speech that had just gotten the student preacher fired, the one about hating Jews and Christians. "Don't you see how that kind of rhetoric fuels the hatred that results in the acts of terrorism that you're condemning?" I gave him credit for listening to me. That was more than the leaders at my mosque did with me. But I didn't change his mind.

"That's in the Qur'an," he insisted.

But it wasn't. It was just in his interpretation of the Qur'an.

This was all an odd turn of events for me, but not surprising. I was on a loop between Mecca and Morgantown. I went to the holy city of Islam from my hometown with the markings of my past in the West—my son and my attitudes. I returned to Morgantown rich with the experience of the hajj. In Morgantown I tried to implement many of the teachings I learned on the hajj, becoming a more awakened and empowered woman in Islam. And in doing so I found myself back in Mecca, uncovering the intricate system by which Saudi Wahhabi clerics are able to disseminate the hate that fuels acts of violence by Muslim extremists.

While denouncing violence against civilians, Sheikh al-Ghazzawi was proud of his role in spreading the gospel, so to speak. He said that his site got three thousand hits per day, of which about one-third came from the United States. Oregon, where al-Ghazzawi had friends, had about fifty hits a day. The number of "unique visitors" from Virginia hit the hundreds on Wednesdays and Thursdays, when religious leaders were preparing their Friday sermons. He said the site served seven thousand Muslim clerics worldwide in seventy countries.

To some critics who have quietly discovered him, al-Ghazzawi is the Goebbels of Hitler Germany. As an empowered woman in my religion, I had found him out, and I was going to go back to the government of Saudi Arabia to hold it accountable for the rhetoric of hatred spewing out from the holiest place in the Muslim world, the city for which the Saudis had declared themselves custodians, contradicting not only their advertisements in the West but, more importantly, the very principles of Islam.

I tried to get comment from the Saudi government but wasn't able to get a statement on the sermons.

What scared me was the ideological loyalty garnered by this type of rhetoric. To my surprise, I got a call from the Egyptian research assistant professor who had called my father an idiot and told me that nobody at the mosque respected me. He had missed the Friday sermon because he was in Washington, D.C., making a presentation to the FBI. He was the researcher on a National Science Foundation grant at West Virginia University looking into dental forensics; the FBI was a collaborator. What he said next disturbed me. "I read what was said. There was nothing wrong with it religiously. It is just that Islam says that we have to be diplomatic. That sermon could have been said in Egypt but not in America." I was stunned.

"There was *everything* wrong with the sermon," I said, "and it absolutely does *not* represent Islam." Another fact bugged me: I knew about the importance of dental forensics in identifying the victims of terrorism. In Mariane Pearl's ninth month of pregnancy I got a request from the FBI to try to find Danny's dental records to identify his remains. In the contradiction of Saudi Arabia itself, how could a man who believed this way be working in America? This was the same contradiction that puritanical Muslims in the West expressed when they held sexist attitudes but freely participated in American society. If we were going to truly live up to the ideals of an honest society, it seemed to me, Muslims had to reconcile these contradictions, guided by the principles of tolerance and equity.

HAPPINESS CLAIMED

PARIS—Like many in America, I became engrossed in the national bestseller *The Da Vinci Code*, set at the Louvre in the City of Love. I hadn't read any mysteries since my days devouring *Nancy Drew*, but my friend Pam Norick handed it to me as she was reading an early draft of my book. "This will resonate with you," she said.

She was right. The story of a search by a Parisian cryptographer and an American art historian for the Holy Grail of Christianity was actually a parable for a universal search for the sacred secrets of women's power hidden under centuries of rewritten history. "The Holy Grail represents the sacred feminine and the goddess, which of course has now been lost, virtually eliminated by the Church," the historian tells his partner.

> The power of the female and her ability to produce life was once very sacred, but it posed a threat to the rise of the predominately male Church, and so the sacred feminine was demonized and called unclean. It was *man*, not God, who created the concept of "original sin," whereby Eve tasted of the apple and caused the downfall of the human race. Woman, once the sacred giver of life, was now the enemy.

In the margin of the book, I wrote just four words: "Just like in Islam." The Qur'an doesn't even say Eve was to blame for Adam's exile from paradise. But somehow, centuries later, the men at my mosque told me that I, as a woman, had to stay physically apart from them or I would corrupt them.

It had been in Paris in the spring and summer of 2002 that I had first battled with the doctrine from my Muslim world that judged me as impure, unclean, and illegitimate for having conceived a baby out of wedlock. It was the season of my hell. Two years later I felt as if I had rejected that doctrine. The only way I could really know, however, was to return to Paris, where I had spent many hours weeping over my shame and sense of worthlessness. I had gone to Paris with Danny's wife, Mariane, after we left Karachi, still trying to understand what had happened in Pakistan. We lived in a pied-à-terre that Mariane and Danny had bought when they moved to Bombay for Danny's assignment there. It sat in the charming neighborhood of Montmartre. Patiently, we waited there for Danny and Mariane's son, Adam, to be born.

In my fourth month I lay doubled over with pain. I had just heard that Danny's kidnappers had wounded him in the stomach during his captivity. My insides were already tight from the emotional anguish that my son's biological father was causing me. I was a Muslim woman struggling to simply emerge from the darkness into which I had descended. That night I wept from a deep, stabbing pain in my abdomen.

"My baby! My baby!" I cried into the darkness. Mariane emerged from her bed in the other room, her eyes wide, her response steady. Her friend Ben was waiting for a phone call to drive her to the hospital for her delivery. Instead, it was me who had to be rushed to the hospital. The French doctor at the hospital examined me carefully. He talked to me gently about the external anguish I was internalizing. He assured me that my unborn baby was safe. "Your baby is protected within your womb," he told me.

The womb. It is *rahim* in Arabic, and its derivative, Al Raheem, is one of ninety-nine names for God in Arabic, meaning "the Compassionate." I learned in that moment that I had to be compassionate toward myself. I had to forgive myself for the errors in judgment I thought I had made, the wrong assumptions that plagued me, the disappointment that I felt I had become. With the evidence of my healthy baby in my womb, God was being compassionate toward me.

So, two years later, I returned with my son, healthy and joyful, to Paris and walked through the doors into that pied-à-terre where I had lived with my quiet torment. The room was filled this time with the chattering of Shibli and Adam, who were learning to say "thank you" in every language I knew. "Thank you," I said. Adam responded in kind. Shibli smiled: "Thunk you."

"Shukran," I said in Arabic.

"Chukrun," said Adam. "Chukrun," said Shibli.

"Shukriya," I said in Urdu, with a lilt at the end.

"Shukheeeehah," said Adam, with the lilt just perfect. "Shuheehah," said Shibli, also with an inflection just right.

As day turned to night, my son's breath filled the quiet of the night like the pulsation of the divine that is universal and timeless. The next day, in an outing with Adam, Shibli and I went together to the Mosque of Paris for the Friday prayer. "Allahu Akbar," I told Shibli, as we drove in a Paris taxi. "Abababoo," he answered.

Inside the mosque the women were crammed into one arm of a verandah that lined an inner courtyard. When we tried to enter one door, the men shepherded us to another door. "Femme. Femme," a man shouted, directing us to a narrow passageway for women, beside a grander entry for men. In any language, his words were the same. Woman. Woman.

Aspects of society try to corral us and subordinate us. During prayer I stood in a row behind the men with Shibli in my arms, close enough to read the size on the rear label of one man's Levi 501 jeans and the "Algérie" on the bottom of a boy's socks. When the prayer ended, the men rushed through even our narrow entrance, and the ushers gestured for the women to wait. I wanted to get outside to rejoin Adam, who was with the babysitter. When I kept trying to leave, the usher kept gesturing me away. I thought of Mecca, where we piled through the gates without having to wait for men to pass. I decided not to wait any longer. When I finally insisted on leaving, the men made way for Shibli and me.

Muslim women—indeed, all women—still have so far to go in our world to reach the status the divine gives us. But somehow in each of our lives we can claim our intrinsic human rights to self-determination and happiness.

In Paris I was in a celebratory mood. We raced through the exhibits with their looming elephants at the Natural History Museum and then strolled through the Jardin des Fleurs. I felt triumphant and happy, enjoying life with my son and knowing that we were making strides within the Muslim world toward the full realization of women's rights. I darted through the grounds with Shibli and Adam in the Jardin des Fleurs and spun around on the merry-go-round with them. As I pointed out blooming bursts of color to Shibli, my exclamations were simple. "Allah!"

These beautiful flowers were expressions of the divine. We need to revel in the good and the beautiful and transform the ugly. This is how I want Shibli to know God—in the beauty of the world. Shibli answered me each time: "Allah! Allah!"

At home in Montmartre, I danced with Shibli in the quiet of our Friday evening, a string of paper rosebuds lighting the room like Christmas lights. Twirling him in my arms, I sang to him in Urdu, "My son, my son, you have come, you have come." As he lay down, he echoed my words to lull himself to sleep. I drew him close to me. And then I released him. Such was the embrace that I felt the divine had given me. I went to the holiest cities of my religion a broken woman. Through the process of transformation that was my hajj, I was now a woman with a deep sense of place and purpose.

Breathing in, I felt immensely different from the woman who had sat there two years earlier. I felt free.

NEW TRIUMPHS

MORGANTOWN—Back in the United States, the Islamic Society of North America was holding its annual convention again in the Windy City. The year before, its pilgrimage tour had opened my eyes to how a woman could be given full and equal access to public space in Muslim society. And its annual convention had shown me women at the dais as scholars, filmmakers, and rap poets.

That was why, early in the year, I filled out a proposal for a presentation from the article my father and I had written for the *Journal of Islamic*

Law and Society on the issue of women and mosques. My father rushed to his office to fax it before a midnight deadline. He didn't get done until 3:00 A.M., because of technical difficulties only my father could find, but to my delight, my proposal was accepted, albeit for a ten-minute presentation on a panel I would share with others. I was honored to get even a minute, I had to admit.

Since I'd sent the proposal, I had had a touchy relationship with the organization. After I filed the police report against the men at my mosque, the society's secretary general, Sayyid Syeed, had been fatherly when I turned to him for help. But its leadership had criticized me after I sought its mediation on the takeover of our mosque and the intolerant speeches preached at the pulpit. In the *New York Times* essay, I had accurately said that the society prefers to stay out of local community disputes. The society responded by running an article in its magazine accusing me of launching a "smear campaign" against the organization. It claimed that I had promised to keep its name out of the essay. A professional journalist for fifteen years, I would never make such a pledge. I considered withdrawing from the conference, but I knew I couldn't. I had to bring the message of the denial of women's Islamic rights to the convention so that as a community we could deal with the problem. I also knew that my presentation would be more like a ten-minute parachute infiltration. I would be an insurgent. I prepared for the worst.

I pulled out a document I had started writing eight months earlier after my article about women's rights in mosques was published in the *Washington Post* and I began receiving support from similarly disenfranchised women in Muslim communities around the world. I called it "An Islamic Bill of Rights for Women in Mosques." This document boiled down to ten simple rights, starting with women's right to enter a mosque, a right that is denied throughout so much of the Muslim world. I sent a draft to my friend Saleemah in Atlanta with the note: "The revolution has begun!"

Over the months, I had vetted the bill of rights with the Martin and Martina Luthers of Islam—Dr. Alan Godlas, Dr. Kecia Ali, Dr. Omid Safi, Ahmed Nassef, Dr. Mohja Kahf, Dr. Aminah McCloud, and Dr. Amina Wadud. Omid suggested that I didn't have to rationalize every right with a hadith. Dr. McCloud suggested that I distribute the rights on bookmarks. I called a husband-wife graphic design team in Morgantown. I thought it was appropriate, in the spirit of pluralism, that the husband's first name was Christian. (The wife was named Paige.) Their design studio was called New Life Arts, and that was what I was trying to do: breathe new life into our communities. In a day they churned out a design with the verse from the Qur'an that inspired us to "stand out firmly for justice" even if it meant testifying against our own kin. They enlarged the book-mark to a poster I could use in my presentation. "It's ready," they said, calling me to meet them at the Blue Moose Cafe, a coffee shop in down-town Morgantown. When I arrived, I gazed admiringly at the work. "Our cause is for real."

Shibli looked at the drawing of the mosque that Christian had sketched and offered his approval: "Ababooboo."

When it came time to carry the poster board and bookmarks to the car, I had to navigate past young university students lined up to go into a bar notorious for serving alcohol to minors under the legal drinking age of twenty-one. Outside stood one of my greatest detractors from the mosque. He owned the building that housed the club. His name used to be on the license when the club was operating as Speedy's, but it had been fined for serving alcohol to minors and reopened under a new name and a new owner, an American guitarist in a local band. I never saw the guitarist around, but I regularly saw the man from my mosque and his brother at the club, accepting cases of beer from Budweiser delivery men and others. His brother was a young man who came to Friday prayers in a luxury con-vertible and quietly slipped in and out of the mosque in a traditional gown. That night he was dressed in jeans and a T-shirt, and the women lined up outside the bar were wearing micro-miniskirts. According to sharia, the letter I would have to wear, if I were Hester Prynne, would be "Z" for zina. Ironically, the name of the bar was Club Z.

Appropriately, I had run into a former professor at the Blue Moose. I had taken a course in international human rights from Jim Friedberg about twenty years earlier as an undergraduate honors student. A graduate of Harvard Law School and a Jewish-American, Jim had spent his career studying international law and human rights. In his class I had learned the concept of self-determination and the value of the human rights intrinsic to the lives of all people. I had borrowed the concept of self-determination to create the Islamic Bill of Rights for Women in Mosques. As we walked past Club Z, I was grateful that I had evolved from being a silent participant in his classroom to a revolutionary on the streets of Morgantown, passing by my fiercest detractors with bookmarks and a poster reclaiming my Islamic rights as a Muslim woman.

THE ISLAMIC BILL OF RIGHTS FOR
WOMEN IN MOSQUES

CHICAGO—Going to the Windy City meant going full circle. I had left in 1992 to pursue a lie. I had walked out of a relationship with a man because he was a Christian. I married another man because he had the right pedigree: Muslim. In the twelve years since, I had tried to resolve the paradoxes within my identity so that I could live truthfully and sincerely.

I was committed to being honest about who I am. Most women, although not all, wore the hijab in Chicago. Even women who didn't ordinarily cover their hair did for the convention so that they wouldn't be the subject of gossip. I cover my hair only in the mosque, and I wasn't going to do it now just for public appearance.

After all of the other panelists had spoken—most with Power Point presentations—I took the podium. I gazed softly at the audience and thanked the Islamic Society of North America. I explained that the presentation was the result of almost two years of work inspired by the transformative experience of praying together with my family in Mecca on the holy pilgrimage of the hajj in February 2003. I had made that journey with the help of the Islamic Society of North America, and I thanked the society for that experience and the opportunity to speak at the convention. My points were simple. "Islam is at a crossroads much like the place where the prophet Muhammad found himself when he was on the cusp of a new dawn with his migration to Medina from Mecca. Medina became 'the City of Illumination' because of the wisdom with which the prophet nurtured his ummah. In much the same way, the Muslim world has the opportunity to rise to a place of deep and sincere enlightenment, inspired by the greatest teachings of Islam. It is our choice which path we take. It is our mandate to take action to ensure that we define our communities as tolerant, inclusive, and compassionate places that value and inspire all within our fold."

The problem was clear. "There are many model mosques that affirm women's rights. Yet women are systematically denied rights that Islam granted them in the seventh century in mosques throughout America. Islam grants all people inalienable rights to respect, dignity, participation, leadership, voice, knowledge, and worship. These rights must be granted to women, as well as men, in the mosques and Islamic centers that are a part of our Muslim communities. Islamic teaching seeks expressions of modesty between men and women. But many mosques in America and beyond have gone well beyond that principle by defining themselves with cultural traditions that perpetuate a system of separate accommodations that provides women with wholly unequal services for prayer and education. And yet, excluding women ignores the rights the prophet Muhammad gave them in the seventh century when he created a Muslim ummah in Medina and represents innovations that emerged after the prophet died."

I gave evidence of the rights denied in mosques throughout America and laid out the Islamic arguments that had empowered me to take action in my mosque in Morgantown. "It is time for our communities to embody the essential principles of equity, tolerance, and inclusion within Islam," I said. "And it is incumbent upon each of us as Muslims to stand up for those principles."

I told them what I had come to realize in the two years since January 2001 when the Dalai Lama had set me on my path toward Mecca. Terrorists transformed our world into a more dangerous place when they attacked the World Trade Center and the Pentagon on September 11, 2001. Before we knew it, a minority of Islamic fundamentalists who preached hatred of the West were defining Islam in the world. Alas, moderates, including myself, have been a "silent majority," remaining largely quiet. A combination of fear, shame, and apathy has contributed to a culture of silence among even those of us who are discontented with the status quo in Muslim society. Moderate Muslims have a great responsibility to define Islam and their communities in the world. For me, this effort started at home when I walked up to the front door of my mosque for the first time on the eve of Ramadan 2003. It is time, I said, for us to reclaim the rights Islam granted to women in the seventh century. Toward that end, I humbly introduced my poster with the Islamic Bill of Rights for Women in Mosques.

The rights are simple: the right to enter a mosque; the right to use the main door; the right to have visual and auditory access to the *musalla* (the main sanctuary); the right to pray in the main sanctuary without being separated by a barrier; the right to address any and all members of the congregation; the right to hold leadership positions, including positions on the board of directors; the right to be greeted and addressed cordially; and the right to receive respectful treatment and to be exempt from gossip and slander.

After reading the rights, I told the audience, "Ultimately, it is incumbent upon Islamic organizations, community leaders, academics, and mosques to respond to this call for improved rights for women in mosques by endorsing and promoting a campaign, modeling it after their very suc-

cessful educational and legal campaigns to protect the civil liberties of
Muslim men and women in other areas. To do so would honor not only
Muslim women but also Islam. The journey is never complete, and a long
road remains in front of us, but we have as inspiration a time in the seventh century when a new day lay ahead of a caravan trader who had as
much to fear as we do today but nonetheless transcended his doubts and
fears to create an ummah to which we all belong today. Allow us all to
rise to our highest potential."

With a deep breath, I sat down, not knowing what to expect next.

Although there were four other speakers, a torrent of questions came at
me when members of the audience stood at the microphone.

There were three hecklers. One admonished me for not saying the
code phrase "Peace be upon him" after the name of the prophet. Another
part of our inside language is "Sall-Allahu aleyhi wa sallam" (May the
peace and blessings of Allah be upon him, abbreviated as SAW), said after
any mention of the prophet or an angel. "The Clans" in the Qur'an
(33:56) says, "The Prophet is blessed by God and His angels. Bless him
then, you that are true believers, and greet him with a worthy salutation."

At the dais, the director of the Long Island mosque, Faroque Khan, a
physician originally from India, had just spoken about the powerful inter-
faith work his mosque had done after 9/11 by opening its doors, and he defended me from his seat. "She is a brave daughter of Islam. Do not criticize
her for such little things." The critics were undeterred. A young man stood
up and identified himself as a member of the Muslim Students' Association.
"Where is your proof?" he demanded angrily, shaking his head, his beard a
blur in front of me. I pointed to the seventy-four footnotes in the reprint of
the article my father and I wrote for the *Journal of Islamic Law and Society*.
"The Sunnah of the prophet will never change," he said, shaking his head
fiercely again. I stared at his eyes, so wide and menacing. *I will never forget
those eyes*, I told myself, not realizing how useful that observation would become when I confronted the young man's rage again, days later.

At that moment, though, I didn't know I'd ever cross paths with him
again, and I actually felt sorry for him that he felt so threatened by the
simple bill of rights. I wanted to scream: these rights *are* the Sunnah of
the prophet. I knew what lay beneath his anger. Some men don't want to
relinquish the power and control it has taken them centuries to accumu-
late. Some men think it is their God-given right to express this power and
control over women. But the prophet gave women rights that men deny
them today, and it is our Islamic duty to reclaim those rights so that we
can be stronger citizens of the world.

y-four-year-old African American woman from Boston, Nakia
od up. The women in her mosque prayed in a urine-stained,
room that doubled as a storage closet. And they accepted the
status quo. "I feel so alone. What advice do you have for someone like
me?" she asked, her voice trembling.

"You are not alone," I told her. "So often I have stood physically alone
in my mosque in Morgantown. But I have felt the spiritual press of so
many kindred spirits who stand with me. I am with you. You are not
alone."

Afterward, I was mobbed. I hugged so many women, young and old,
that I lost count. And I received the encouragement of so many men,
young and old, that my faith was renewed. "We did it!" I told my parents
when I called home later.

THE PROFESSOR

Two roads diverged in a wood, and I—
I took the one less traveled by,
And that has made all the difference.
                    Robert Frost (1874–1963)

LOS ANGELES—"The professor would like to invite you, your mother, and
Shibli to visit him at his home," said Naheed Fakoor, an Afghan American woman who was the assistant extraordinaire to UCLA law professor
Khaled Abou El Fadl. She mentioned delicately that he rarely invited
anyone to his home. I understood the power of this invitation. In my estimation, we were getting an audience with the pope of tolerant Islam.

A year earlier I had not even known that this professor existed. It had
taken me months to memorize all the syllables of his name and the order
in which they are said. More often than not, I referred to him as "Abou
Khaled . . . oh, you know, the UCLA professor of law." My kindred spirits
within the Muslim world knew of whom I spoke. When I told Dr. Alan
Godlas, the Islamic studies professor at the University of Georgia, that I
was going to be meeting "the Professor," as Naheed allowed me to call
him, Dr. Godlas got excited. "He is the best hope for Islam in America,"
he said. Not only could the Professor speak to disenfranchised Muslims
like me, but he had the grounding in Islamic jurisprudence and original
texts to be able to communicate with the mainstream puritanical set who

were in positions of authority in our communities. His students and friends had created a website for him called scholarofthehouse.org, inspired by an award he received by that name when he was a student at Yale University. They called him "the most important and influential Islamic thinker in the modern age."

How had I missed him? Like liberal movements everywhere, he wasn't well funded, and though he'd published extensively, he couldn't compete with the publishing machinery that he opposed intellectually and ideologically—the Wahhabi brain trust coming out of Saudi Arabia. Until the summer of 2004, he was a visiting professor at Yale Law School; he had just returned home to California, where he was a professor at the UCLA School of Law. He sprang from American academia, earning a BA at Yale, a JD at the University of Pennsylvania Law School, and an MA and PhD from Princeton.

What separated Khaled Abou El Fadl from many intellectuals was his training in Egypt and Kuwait in Islamic jurisprudence: he was a high-ranking sheikh. I didn't know about him until a friend sent me a copy of a book he had written, *The Place of Tolerance in Islam*. My mother read it first. When she was finished, she closed its cover with a sigh of relief. "Since my childhood days," she said, "I was told that only Muslims would go to heaven and no others. I would ask, 'Why?' We were born Muslim by accident. Why should others be denied heaven because they were born Christian, Jew, Buddhist, or Hindu? Nobody would answer me. They would tell me not to ask such stupid questions. For the first time, someone has answered my question and confirmed my belief that this assumption is wrong and that, in fact, the doors of heaven are open to all good people." She paused. "Khaled Abou El Fadl is a great man," my mother said. "He is the first person who helped me understand and *believe* in Islam." When I had called him to seek his guidance on the trial that I faced, I told him my mother's story. "Al-hamdulillah," he said, simply. Praise be to Allah.

It was enough to speak to him by phone. It was an honor to be beckoned for an audience. My response surprised even me. After all, I had met with senators, celebrities, and heads of state. What was a *professor*? But I knew I had to make this trip. I just didn't know why.

In Los Angeles I framed the Islamic Bill of Rights for Women in Mosques for the Professor and wrote below it two words: "In Gratitude." I dressed Shibli in an ornate silk vest from India. I brought a copy of *Time* magazine with an essay in which I'd written about the struggle for the soul of Islam and an article in which the Professor argued against the theological

logic that al-Qaeda leaders such as Abu Mousab al-Zarqawi use to sanction the beheading of prisoners. Since Danny's beheading, militants from
Iraq to Saudi Arabia had turned to this execution-style brutality to kill
hostages. The Professor told *Time*, "Al-Zarqawi searches for the trash that
everyone threw out centuries ago and declares the trash to be Islam." His
words resonated with me on so many levels, including in the battle to win
rights for women in our Muslim world. Having some idea that the meeting would certainly be meaningful, but not knowing quite how, I departed
for his home with curious anticipation.

After winding slowly through rush-hour traffic, we arrived at the Professor's house at the corner of a street in a Los Angeles suburb. It was surrounded by a security fence, a reminder of the danger in which he lived.
He had taken on Wahhabi ideology with frontal attacks on their theology
as flawed and un-Islamic. Even Muslim American organizations lashed
out at him when he penned an op-ed for the *Los Angeles Times* after 9/11,
criticizing Muslim leaders for not condemning the attacks.

The Professor's assistant, Naheed, warmly beckoned us through the
gate. His wife, a gentle and beautiful woman by the appropriate name of
Grace, welcomed us at the door with a smile and embrace. The door
opened into rooms that swept into each other. The walls were lined from
floor to ceiling with one thing: books. The books in some sets were lined
up next to each other so that their bindings spelled words in Arabic.
Shibli climbed up the stairs, ready to explore without a guide, and gazed
upward at a wall tapestry of pilgrims circling the Ka'bah. He absorbed the
familiar scene and exclaimed, "Ababooboo." Appropriately, it was time
for the sunset prayer.

The Professor emerged in a flowing black cape, under which an embroidered collar peeked out from the traditional Arab gown he was wearing. What struck me immediately was his physical vulnerability. He
leaned on a cane and walked smoothly but slowly with small steps, extending his hand toward me gently. This in itself was significant. Saudi
scholars had ruled that a man shaking hands with a woman was "evil" and
haram, or unlawful. I took his hand gently. "Thank you for the honor of
this invitation to visit you," I said. Not close enough to extend her hand
to the Professor, my mother raised her cupped right hand to her forehead
in a high-browed Indian Muslim gesture of respect between men and
women. "Adhab," she said, in an Urdu greeting.

The Professor immediately lent me his support in the trial that my
mosque had started against me. He reveled in the spirit of the Islamic Bill

of Rights for Women in Mosques. "This is good," he said. And he cheered the writing that I was doing.

When it was time for prayer, the Professor did another remarkable thing. He prayed with his son on his left side and his wife on his right side. Naheed, my mother, and I lined up behind them. He needed help going into prostration and then standing up again, but his spatial arrangement was more than just practical. He believed in the intrinsic right of women to stand on a par with men.

In prayer at the Professor's house, I felt free for the first time in so long. Even though I was physically behind the Professor, I did not feel disrespected. Many Muslim men say they are expressing respect for women in their desire to protect them through segregation. But I knew the Professor didn't want to silence me. We sat around a circular dining table until late into the night, and he honored me when he revealed that he had read my first book and supported my voice. "It is a great victory that you are writing. You can only testify to the truth that you know," he said. "And you are doing that." He spoke with candor, even using the word *sucks*, and he lamented that Muslim leaders ran their communities as if they were playing Monopoly, collecting properties and building symbolic hotels of power.

The next night we returned, and my mother, Naheed, the Professor's wife, and I prayed in the same row as the Professor and his son. When we broke from prayer, the Professor led us through a tour of his library. To call it a library is an understatement. His books filled every wall in his room. ("Does he sell them?" my mother asked afterward, perplexed about why he collected so many books. "Mom!" I admonished her, but privately I could understand her curiosity.) He pored through Arabic texts to show me the works of the two thousand women jurists Islam has had since its inception in the seventh century. A year earlier I couldn't even name the century in which Islam was born. Now I knew the number of women jurists we've had. I was both amazed and astonished. On these shelves was the secret history of feminine power that centuries of male domination had erased. He pulled down a book published by a *madhab*, or school of jurisprudence, that had been destroyed. Both the school and its prayer had been led by a woman, he told me. "What?" I exclaimed. "A woman?"

"Yes, a woman."

I was surprised to see a familiar book on his shelves—my first book, *Tantrika*. And it was the hardcover edition with the image of a woman's bare torso on the front. Some Muslims had protested the cover when

editor Ahmed Nassef put it on the Muslim WakeUp! website. I had put a
yellow Post-It over the torso when I took the book to the Islamic Society
of North America convention. But here it was, uncensored, beside the
works of the great Muslim jurists. And to my shock, the Professor asked
me to autograph the book. So many men—and even some women—
within my religion had discounted me and discredited me because I wrote
truthfully about the most intimate challenges in my life. But the Professor
did not make me feel ashamed. Instead, he affirmed me.

I handed the book back to him with an inscription that couldn't capture the gift he had given me with the respect he had shown for my intellect, voice, and being. As he stood, he lost his balance and his cane
slipped from his grip. My heart fell as I witnessed the physical vulnerability that accompanied the Professor's spiritual fortitude. Grace moved
swiftly to help him regain his balance. Without pause, he looked me in
the eye and took my hand, gently, in parting.

"Asra, do not let anyone deter you. Continue to be courageous," he
said clearly. "You are on the right path."

"Thank you, Professor," I said quietly. Touched by his words and his
sincere gestures of kindness and respect toward my family and me, my
heart wept. He recognized that I had struggled hard to resolve the dissonance between the intimate areas of my life and my religion. So often we
live with guilt about our sexual lives. But I had found a peace with the decisions I'd made about my body, and I had claimed the worthiness of my
spirit. I could embrace Islam.

Religion isn't meant to destroy people. The Professor recognized that I
had struggled to answer the question of who I am and where my faith rests
in my identity as a Muslim woman. I knew he was not speaking just to me
but to all people—women and men, girls and boys—who choose the path
of honesty, justice, tolerance, and compassion. This has been my struggle,
but it is the struggle of all people as well. Every time I've spoken out, there
have been people ready to take me apart. But I've constructed my own
identity, and my son is a vital part of it.

Shibli was by now in only his T-shirt. I had taken his President
Musharraf jacket off long before as I struggled to get him to sleep. He flew
toward an ornate room divider, ready to knock it down. Stopped mid-
stream, he bounded over to a rescued blind dog. Afraid for its safety in the
company of an energetic twenty-month-old, I scooped Shibli into my
arms and bade farewell to the Professor and his wife. From what I could
see, he was as close as Islam comes to having a Dalai Lama who preaches

peace and tolerance. It was fitting that I should end one phase of my journey with him and, emboldened, set off from there to move forward.

The next night I found myself in the most unexpected of places—a Democratic fund-raiser for presidential candidate John Kerry. Screenwriters, producers, and agents spilled through a sprawling home in Hollywood for the event. Only in America, it seemed, could I go from the home of the pope of tolerant Islam to the front row at a backyard concert by African American pop star Macy Gray. When she took the stage, her Afro seemed more like a halo. In a moment of abandon and glory, she yelled to us in the audience, "Look up at the heavens! And yell to the heavens with the glory of you!"

I yelled to the heavens with the glory of me. I could feel the spirit of Hajar in the twinkle of the stars above me, just as I had felt her spirit when I walked in her footsteps in Mecca between the hills of Safa and Marwah.

DEATH THREATS AND THE FINAL FRUITS OF PILGRIMAGE

Only a free individual can make a discovery.
Albert Einstein (1879–1955)

MORGANTOWN—After our plane from Los Angeles touched down, I was afraid my father would have a heart attack as he nosed our Chrysler minivan south on Interstate 79 to Morgantown. My brother and he had just picked up my mother, Shibli, and me from the Pittsburgh airport. He was agitated about a comment posted on Muslim WakeUp! by a person who didn't appreciate my Islamic Bill of Rights for Women in Mosques. The critic had posted a note under the subtle ID of "deathtoasra" and the false e-mail address dta@dtasra.gov.saudi, trying to suggest that the message had originated from the government of Saudi Arabia.

> finally some people are getting an idea of what this trash is prmoting—read the works of michael muhammad knight, available on this site and online, where he implies that he has slept with asra, and says that he HAS Witnessed Asra praying in the SAME LINE WITH

MEN, and heres the fun part, BEING LED BY A FEMALE IMAM.
Man wallahi [my God] if these jackasses go to Saudi, they'll get their
heads cut off—as they SHould To have men and women pray in the
same line led by a female imam is a SLANDER on The entire Quran
and Sunnah and a CURSE TO ALLAH. Even for ADOVCATING
such Blasphemy, One is likely to get killed in most places in the Middle East, and not by the goverment, but by private citizens outraged at
this shit. One of these days, she's going to get herself killed, and the
fun part is going to be when they try to give her an islamic funeral and
bury her with the muslims—lol [laugh out loud] cant wait to see that—
may allah remove these terrorist imposter pigs from this earth and destroy them by the Hands of the Believers, inshaa Allah [God willing]

Even with all of the barbs I'd received, this was the first call for my
death. And over what? Over my support for mixed congregational lines,
as we'd had in Mecca, and women prayer leaders of mixed congregations.
The posting tried to attack the place that so many people considered a
woman's Achilles' heel: my sexual honor, suggesting that I had had sex
with Mike (I had not, and Mike had never written that we had). But in
the repressed world of puritanical Islam, men and women being friends
and, God forbid, praying together was hedonistic in all ways.

Another aggressive posting was also disturbing. It reflected the double
standard in Muslim society, as in so many cultures, toward women. Mike
Knight had written about his own sexuality and in one of his articles had
recounted a conversation in which he told a woman, "I have sex with
girls!" There were no protests about his sex life in the comments to the article. In my article I had mentioned that a few of those who read my article about women's rights in mosques were judgmental about my unwed
pregnancy. The second disturbing posting seized on this line.

. . . As for Nomani, it is a shame no one has taken on the task of
exposing her for who she is. If you read the works of one Micheal
Muhammad Knight, he claims to not only have personal knowledge
of Asra, but asserts that he has witnessed Asra praying side to side
with men, as in feet touching, as in the same row, BEHIND A FE-
MALE IMAM. As this Knight fellow is becoming quite popular and
was invited to speak at ISNA, and since Nomani has not refuted his
claims, this DESTROYS ANY credibility Nomani may ever seek to
have. This is her true agenda: to have men and women praying side to
side behind female imams.

AllahummAhfazni. Oh and if Asra is so in favor of ISlamic Law, why doesnt she appear before a qadi, and accept her stoning? Doesn't the same "Islamic Law" she touts for this women's bill of rights also MANDATE Death by Stoning for any Woman who commits adultery, and confesses, as Asra has done on so many occasions. May Allah destroy the enemies of Islam. If nomani keeps up this facade for long, she might meet the same end as all the other enemies of Islam, either she will self-destruct or some soldier of Allah will destory her. Either way, victory will be Allah's.

My father was clearly disturbed by these e-mails. "Dad, it's okay," I reassured him from the backseat. "It's about power and control. The people protecting the status quo don't want to lose their power and control, so they try to make us afraid of speaking out. Don't worry. I *will* find out who is behind the postings."

With some sleuthing, I traced the threats to a Pakistani American honors undergraduate at Penn State University, just hours from me in Morgantown. He was the bug-eyed young man who had challenged my sources for the bill of rights. I reported him to the FBI, and it opened a criminal investigation. Then I took on the more introspective task of processing the impact of his threats on me.

I had known that the two issues that disturbed the Penn State student—women prayer leaders and mixed congregational lines like the kind in Mecca—were controversial. In fact, he did me a favor: he made me take a position and fully embrace these Islamic rights that scholars, including Dr. Khaled Abou El Fadl, Dr. Amina Wadud, Dr. Omid Safi, and Dr. Aminah McCloud, had told me women have in mosques, namely, the right to lead prayer and the right to freely assemble anywhere, including the front of the mosque and shoulder to shoulder with men.

So often in Muslim society, as in most societies, fear is used to control people. Being American, I have freedom of movement, thought, and voice. Democracy and these freedoms bring me closer to my faith. I didn't know these realizations would come to me as a result of my pilgrimage, but they did. We do not need repression to engage people in the spiritual life. When I confronted the traditions in my Muslim society, I discovered that the gem of Islam is quite pure and good below the layers of repressive sedimentation.

One evening I took Shibli outside and knew that before me stood the fruits of my pilgrimage. He wore a shirt from a clothing label, Mecca, popular among rap artists; it said "Mecca Since Day One" on his sleeve. By traveling with me on my pilgrimage as a baby, Shibli embodied for me the essence of Mecca. My son gazed through the evergreen trees at a waning full moon and smiled and squealed in delight.

"Moon! Allah hu," he said, in the expression of the universal divine breathed in and the exhalation of the internal divine breathed out. I was teaching him to admire the divine in all of creation. God is. The divine is.

"Allah hu," I answered, smiling. "Allah hu."

EPILOGUE

As Muslims in America, we are engaging in *jihad li tajdid al-ruh al-Islami*—
a struggle for the renewal of the soul of Islam. We aren't trying to change
Islam. We are trying to question defective doctrine from a perspective
based on the Qur'an, the traditions of the prophet, and *ijtihad* (critical
thinking). I was fortunate enough to travel to the heart of Islam through
the pilgrimage of the hajj, and I was blessed to come to know the pulse of
the true spirit of Islam. As a result, I was prepared to join the quiet tide of
reform that is very much under way in U.S. Muslim communities. That
movement eschews bigoted, sexist, and intolerant practices that betray
Islam, the prophet Muhammad, and all of the good people who call themselves Muslims.

Almost five hundred years ago, Martin Luther became a symbol of religious reform when he posted the "95 theses" on a church door in Wittenberg, Germany, calling for Pope Leo X to eschew traditions adopted
by the Catholic Church and return to the essentials of Christianity. The
Muslim world's Martina Luthers are currently leading a reform movement
based on the principles of the prophet's first community in the seventh
century. For instance, four twentysomething women attorneys in Chicago
were added to the board of the Downtown Islamic Center when they
complained about inadequate facilities for women in a new mosque design, and the website Muslim WakeUp! organizes monthly meetings nationwide for building tolerant inclusive communities. The inspirations for
these actions are the women who prayed, spoke, led, studied, and debated
in mosques at the time of the prophet Muhammad. When seven Muslim
women gathered in Morgantown and marched on the mosque to support
women's rights everywhere, they evoked the strength of the historical
mother of Islam, Hajar, a single mother who relied on her will and faith
to raise her son in the desert where modern-day Mecca now stands.

Within the Muslim world, some critics try to marginalize those of us
in the reform movement with personal attacks. But the larger Muslim

community is recognizing that the issue of women's rights is not about one woman or even seven women. Our spiritual membership is vast. In Vancouver, Canada, the political candidate and human rights activist Itrath Syed protested when mosque leaders wouldn't let her address a congregation recently, as they had promised; instead, the imam preached a sermon deriding her. In mosques in Washington, D.C., the community leader Sharifa Alkhateeb removed ceiling-to-floor separations, sat in front of partitions, and prayed in spaces that have been restricted to men before her death in the fall of 2004. At a mosque fund-raiser in Chicago, radiologist Ejaz Rahim cast aside the RESERVED sign on a front table set aside for late-arriving men and took a seat. In Sacramento, California, Nasrin Aboobaker helped found the SALAM Center, whose bylaws include the mandate to "empower women."

Those trying to slow us down need to realize that we are a generation that believes that if change will come tomorrow, we should have it today.

Improving the status of women has been the symbolic expression of broader historical reform. We have model mosques in the United States and Canada that understand that. In Sterling, Virginia, at the ADAMS mosque, women walk through the front door, participate in study circles, serve on the board, and share the same prayer hall as men, separated by only a symbolic partition. Mechanical engineer Uzma Rasheed introduced a martial arts class for women there. In Perrysburg, Ohio, the Islamic Center of Greater Toledo elected Chereefe Kadri, a woman, as its president in 2001. In Toronto, a respected Pakistani Canadian Muslim social worker, Shahina Siddiqui, drafted a report on making mosques "sister-friendly" and raised her voice for women's rights in mosques in an essay in the *National Post*. In support, Muslims widely circulated her column via e-mail. For months the Council on American-Islamic Relations, the civil rights group for Muslims, ignored my pleas for help, saying it didn't get involved in Islamic-Islamic relations. After our march on the mosque, the council circulated Shahina Siddiqui's "sister-friendly" paper with the intention of supporting it.

In Chicago imam Abdul Malik Mujahid added a man's voice to the call for women's rights in mosques in a powerful sermon he delivered in the late summer of 2004. A well-respected leader as president of an Islamic media company, Sound Vision, he estimated that nine mosques out of ten in the Chicago area offer a dearth of leadership roles to women, as well as inadequate space, facilities, and services. In a section of his sermon titled "'Men's Islam' or Islam for All," he said, "While sisters are a full part of the community, many mosques are run as though Islam is just for men."

He called for action: "It is time that sisters come together and provide leadership in clearly defining a Muslim women's manifesto for change in mosques in North America." He proposed that Muslim women in North America start a national women's caucus to bring about three reforms to mosques and community centers: having them publicly state that it is not Islamic to deny women entrance; providing women with space without a barrier and an unobstructed view of the prayer leader; requiring that one-third of board seats be filled by women, one-third by men, and one-third by people born in the United States, allowing for more tolerant attitudes about participation by women. He called for such a caucus to give deadlines to mosques to provide space for women and to convene a meeting of Muslim leaders in the United States and Canada to make the issue of women's rights in mosques a priority.

In Morgantown the cause of *islah* (reform, making righteous) was winning. In the time since I had first entered the main hall, women had begun to walk through the front door, pray in the main hall, fill positions of leadership, and guide community activities.

In a real glimmer of hope, I discovered strong Muslim men and women throughout the United States and Canada who were using intellectual reasoning, expressions of dissent, and important courageous actions to fight to liberate Muslim communities from cultural norms that contradict the Islamic principles of tolerance, inclusion, and equality. A male Toronto mosque planner included tampon dispensers in his design for the women's restroom, as well as a private room for women who would usually stay home during their menstrual period. Muslim women from New York to San Francisco chronicled their efforts to remove screens, step in front of one-way mirrors, and raise questions at community meetings. One called herself a "Muslimah rebel." Muslim men chronicled their efforts to equalize participation and accommodations for their wives, sisters, and daughters.

Even the government of Saudi Arabia has realized that it must reform its puritanical society. "Progress. It sounds right in every language," a radio ad for the government recently proclaimed.

Golam Akhter, convener of the Bangladesh-USA Human Rights Coalition, posted an e-mail in a discussion group in which he declared that women's Islamic rights in mosques are consistent with the Universal Declaration of Human Rights: "One sister's dilema and conviction resolved a problem. So long we pretended these problems as minuscuale and insignificant; possibly the true Islam will be exported from the west to east by wars and interpretation of the holy Quran and Hadith, a fresh, by Islamic scholars from the west."

Averse to the theological quagmire implicit in the use of the word "reform," some Islamic intellectuals, such as Akbar S. Ahmed, Ibn Khaldun Chair of Islamic Studies at American University, are speaking of the movement as a "renaissance." Dr. Ahmed had started a public dialogue with Judea Pearl, the father of my friend Danny, to transform modern-day Muslim relations with the Jewish world through tolerance. He believes in dismantling barriers in the Muslim world that amount to bigotry *and* sexism. On the Royal Jordanian flight to the hajj, I had read an article by Dr. Ahmed, who is also a former ambassador from Pakistan to the United Kingdom. As I was about to be, he had been moved by doing the hajj together with his family, without barriers.

On Muslim WakeUp's comment board, the literature professor Mohja Kahf posted an e-mail from the Islamic studies scholar Amina Wadud, who argued for Meccan-style praying while maintaining separate areas for men and women who want to be segregated. "As far as my opinion on the matter of gender restrictive practices as they are observed in most mosques in the United States and elsewhere," Wadud said, "I am advocating a complete dismantling of gender stratification so that women and men can pray side by side if they want, be led in that prayer by a Muslim man or woman, listen with respect to a *kutbah* given by a man or a woman."

It's clear that certain traditions and ideologies ultimately contradict essential teachings of Islam as a religion of peace. The veteran *Washington Post* Middle East correspondent Carlyle Murphy notes: "The task of reinterpreting Islam for modern times is the essence of Islam's contemporary revival . . . around the globe. If this 'interpretive imperative,' as some Muslims call it, is fully embraced, Islam's revival will become Islam's renaissance, ushering in a new era of intellectual creativity for Muslims. This is the promise of Islam, and the source of passion for Islam."

I believe there are some fundamental changes the world of Islam must make in order to be true to the spirit of the religion. First, we must live by the golden rule common to all of the religions and philosophies of the world. We must respect others. Second, we must open the doors of Islam. Saudi Arabia must open the doors of Mecca and Medina to those who are not Muslim. Muslims around the world must open the doors of their mosques to women and those who are not Muslim. Third, we must open the doors of ijtihad in the Muslim world. Fourth, and finally, we must honor and respect the voices and rights of *all* people.

In my hajj group there were examples of Muslim women thriving in the changes that were transforming our communities and bringing them more in line with the ideals set forth by the prophet Muhammad. The college student who made space in the women's tent in Mina so that my mother and I could sleep with Safiyyah came back from the hajj and rose to become president of the Muslim Students' Association at Penn State and a member of the Vice President's Cabinet of Student Leaders. She graduated in May 2004 with honors in biobehavioral health, planning to pursue a master's degree in public health and social work. And her mother's prayers on pilgrimage came true. In a traditional ceremony in the Harrisburg, Pennsylvania, Hilton in June 2004, she married a handsome Penn State student she had gotten engaged to after the hajj, following a courtship as coleaders of the Penn State Muslim Students' Association.

For the photos she bowed to tradition: in the demure images she is veiled, with eyes cast down. But in others she is beaming openly with a wide smile. In hijab, her mother beams just as widely. Most of the women guests wore hijab, but not all of them. The wedding invitation included the date from the Islamic calendar and a Qur'anic verse (30:21): "And among His signs is that He created mates for you from yourselves, that you may dwell in tranquility with them." Their e-guestbook included congratulations from friends with names like Lisa, George, and Francis, as well as a saying that reminded them that the prophet exhorted men to treat their wives well. "The best of the human beings are the ones [with] good manner[s], and the best ones of those are the ones who are good to their wives." A former Penn State classmate wrote to the couple: "I sincerely pray that Allah blesses your marriage and your future together and truly makes you a garment for one another insha'Allah [God willing]. Oh and don't forget. . . . 'BREAKITDOWNNOW . . . DA NA NA NA!' lol :)."

The triumph of our young hajj friend was the victory of the American Muslim girl next door. It was a tale told more smoothly than my own, it seemed, but I didn't envy her. I was thrilled for her.

Admittedly, redefining boundaries comes at a cost. I've wept from public phone booths in India. Cultural expectations made me want to literally kill myself from the shame of carrying a baby without a ring on my finger. I wept for hours after a mosque leader in Morgantown threatened to issue a restraining order against me just because I dared to sit behind the men during a Friday night study session. And the personal attacks against me for my choice to be honest about Shibli's conception, even though they come from an isolated minority, have hurt me to the core.

In the same way, redefining boundaries also seems to come with costs to society. Social order breaks down somewhat for a while as people live outside traditional boundaries and deal with public disclosure of unwed pregnancies, abusive marriages, exploitation, and other societal challenges. But it is certainly not my intention to dispense with social order as we redefine boundaries. I believe that if rigidity breeds rebellion, hedonism breeds irresponsibility. As we redefine boundaries, it's important to remember that all societies need to be defined as well by personal responsibility.

I believe that society is better served by redefining its boundaries in the spirit of compassion, forgiveness, tolerance, and love that all religions teach. And ultimately, I am convinced, I am a better person as well for having redefined these boundaries. I feel that I am richer and stronger for all that I have gone through. It is not everybody's path that I have followed, but it is the only one I feel I could have traveled. I have suffered some, and I have seen suffering. I have endured betrayal, shame, and hostility. But I have not surrendered, and in my survival and triumph, I am happy.

In the most humble of ways, I know my son directly connects me to God. I went to Mecca in fear, wonder, and doubt. Then I stood in front of the Ka'bah without shame. My body was a vessel for the divine act of creation. Men's interpretations had told me that I was a criminal, but I rejected their judgment. I didn't anticipate the effect that my pilgrimage to Mecca would have on me. After undergoing the most sacred of experiences as a Muslim woman in one of the most repressive regimes in the world, I received a shocking wake-up call when I tried to bring the lessons from the pilgrimage home to my own community in America—one of the most democratic societies in the world.

This book is testimony to the potential for all of us to become empowered, spiritually, intellectually, and emotionally, if we allow it. When I started the book, my goal was to describe my experience doing the hajj. But along the way I found my voice, and the book helped me to clarify my identity as a Muslim woman. What I ended up with was a book in which I've expressed myself in the strongest voice I could muster.

I spent countless days in the downstairs office of a new friend, Ed Jacobs, a professor of counseling at West Virginia University, as he coaxed my voice out of me. He became my life coach, in the vernacular of the new millennium, as I tried to separate my programming from my beliefs so that I could truly succeed as a human being, a writer, and a mother. His message was simple, but one that we so often forget, no matter who we are: "You're doing terrific."

Ed punched a paper cup with a hole and said, "This is your self-esteem." I put another cup inside to seal the hole and put them both, one inside the other, on top of the highest shelf in my writing room at home. I was learning that it was only my own insecurities silencing me—I couldn't blame my religion or my parents—and that it was my responsibility to free myself from the internal wiring I'd inherited from my culture.

Now I hope to amplify the messages in this book with a public voice. I plan to air these messages in a campaign I am calling the Muslim Women's Freedom Tour. I will assert the Islamic Bill of Rights for Women in Mosques and an Islamic Bill of Rights for Women in the Bedroom so that Muslim women's rights are reclaimed throughout the Muslim world. I will do so with the inspiration of centuries of women before me and legions of women around me who have been the Florence Nightingales, Susan B. Anthonys, and Eleanor Roosevelts of the Muslim world, only with names like Rufayda, Aisha, and Khadijah.

Recently, someone sent me a poem titled "The Pious Wife," written by a man, Abu Jameelah. One line reads: "She opens her mouth only to say what is best. Not questioning her husband when he makes a request." In reclaiming the history that has been stolen from us, Muslim women must counter centuries of this sort of programming to stay silent.

These forces are still trying to silence us, as illustrated by the challenges I and other women face at my local mosque. While the moderates, including the first woman elected to office, had won the majority of the seats on the nine-member executive committee in an election in May 2004, they ran into constant roadblocks put up by the puritanical. The battle for the heart of Islam in my community erupted on a bizarre front, but one that could have been predicted: a potluck dinner. The executive committee had voted to hold a mixed-gender potluck dinner, with the kind of intermingling I'd experienced from Mecca to the Chicago Islamic Society of North America convention, but the puritans had protested, ruining the dinner. The president of the mosque proposed banning me from the mosque without trial, and then rallied men to encircle me and intimidate me out of the mosque. In an e-mail from his account at West Virginia University, the professor wrote:

The res traning order is some thing we need to talk about next if we need to go that far. I think if a buch of guys go around Asra and ask to leave if insists on coming, she can be scared off, if she wants to play the intimidation game we will play that also but politly, remindinh her all the time that she is not wanted, I remember even the remarks during Khutba to people praying behind her that their prayers are not accepted used to bother her a lot, now if she reminded all the

time that her existence is not accepted and she is not wanted, we
might not need a restraining order.

The woman member of the executive committee posed a rhetorical
question back to the president: "I wonder if Prophet Muhammad, Peace
and Blessings Be Upon Him, the Messenger of Allah, Subanahu Wa Tal-
lah ["the sacred and the mighty"], would have chosen such a tactic or
procedure to scare off a sister from the masjid?" He didn't respond. Four
members of the board, including the woman, resigned in frustration
around that time. Not long after, there was another confrontation involving the board president who had ordered me to the back door and the balcony a year earlier. He yelled at two women who were trying to meet in
the mosque for a women's study session, ordering them to leave and use
the dingy old mosque. The husband of one of the women arrived, and the
board president repeatedly pushed him, yelling, "Go out!"

The woman who had resigned office wrote to the mosque leaders
lamenting the intolerance toward women at the mosque. She and I had
had a conflicted relationship because, while she had initially supported
my efforts, she allied herself with the conservative men at the mosque to
try to work from within. I respected that decision as the path she wanted
to take and had cheered her election to office. But by the end of the summer she clearly saw the gender barriers that puritanical Muslims put up.
She wrote that perhaps my father had been right and my use of "pen
power" had indeed been essential in tackling men who wanted to deny
women their rights at our mosque in Morgantown. After stepping down,
she wrote: "Part of my resignation was due to absolute and intolerable dis-
gust at these types of men. . . . I needed a break for my own mental sanity
after confronting the dark side of people who claim to be Muslim. . . .
Women are not animals confined to dark corners, to be veiled with voices
at a whisper. We are thinking, intellectual human beings who are going to
stand strong against leadership and men who perpetrate these actions and
thoughts."

The efforts to silence us were many. A professor of psychiatry sent me
an e-mail on the eve of Ramadan about dinners, called *iftar*, for breaking
fasts. He wrote:

Hi Asra,
   I am organizing Iftar dinner every Saturday for families (husbands
and wives) and their minor children (under 18 years). We will be

honored to have your father Dr. Nomani and his wife. We will be honored to have your brother Mustafa, his wife and their children.

To avoid any confusion, I want to make it clear that you are not invited to these dinners.

Thank you.

The noninvitation was effective in throwing a punch to my gut, and it hurt. But I knew we had to persevere. I had been inspired to walk into the main hall of my mosque after I had seen Michael Wolfe's co-produced documentary about the life of the prophet Muhammad. I had sat in the back rows of the Gluck Theater in the student union of West Virginia University. With an invitation by the President's Office of Social Justice, I took to the stage for a presentation during Diversity Week. I presented the Morgantown Model of what to expect when challenging power and control, and I illustrated my points with the cycle of intimidation, emotional abuse, and social isolation that I had seen in the year since my family and I had walked into the main hall. On the exact anniversary of that day, on the eleventh day of Ramadan, the *Daily Athenaeum*, the student newspaper, ran a headline that reflected the victory of our commitment to change: "Morgantown Woman Inspires Change in Muslim World." The story wasn't important because it was about me, but because it was about the reform that we were creating. I recognized that this was what we were doing when months earlier I had plucked the *New York Times* out of the news box in front of the Blue Moose Cafe in downtown Morgantown. Above the fold was a photo of hooded Muslim militants with truck drivers they had taken hostage in Iraq. Below the picture was a headline over another story proclaiming the new reality women are trying to create: "Muslim Women Seeking a Place in the Mosque."

At our mosque, leaders held a gender-segregated party for Eid ul Fitr, the festival marking the end of Ramadan, just like they'd done since my childhood years. "We need to create a new reality," I told the woman who had resigned from office. "I agree," she said. "Islam can be a religion of joy." She bought a jumping pit for children, and organized a multicultural poetry reading for an Eid festival where little girls wouldn't have to write into their journals, as I once did, that they felt like they were in a prison. Our children, in contrast, jumped in joy.

With this success, my father and I drove to a retreat that I had been asked to attend. To my surprise, Daisy Khan—the Muslim leader in New York who had advised me, half in jest but mostly seriously, to make anyone who tried to deny me access to the mosque feel as guilty as possible—had invited me to attend a weekend retreat for "Muslim leaders of tomorrow." I hadn't thought of myself as a leader until our Morgantown mayor, Ron Justice, invited me earlier in the year to speak to his leadership class. I was always the secretary of the math club in high school, never the president. I never thought I was worthy of that role. But Ron told me, "You're an agent of change. You are a leader for progress that is inevitable."

I had come to understand that I had unwittingly become a leader in the place where it is perhaps most important, but also most difficult, for us to show strength—in my own life. I had navigated through contradictions, confusion, doubt, and wonder as I learned to guide myself—and, most importantly, my son—with clarity, truth, and courage. On the banks of the Hudson River, at the retreat, something remarkable happened. Daisy's husband, Imam Fesial Abdul Rauf, had the women pray in a parallel section to the men with women on the right side and men on the left. All of a sudden, I was in the front row. In one year I had gone from the back of the bus to the front.

In a reminder of the challenges still facing us, the president of the mosque sent me a Thanksgiving Day missive with the complaints against me in the trial to banish me from the mosque. The evidence against me: the actions I had taken over the past year to reclaim my rights within Islam. He gave me one week to respond. I took the charges seriously and planned to respond systematically to them. But for the moment I wrote him a simple reply: I'm busy.

While we challenge the status quo, we are busy creating a new reality. At Harvard University in March 2005, Muslim thinkers and activists planned to hold the first convention of the Progressive Muslim Union. Sarah Eltantawi, one of the women who had marched to the Morgantown mosque under the banner of the Daughters of Hajar, is one of the founders. She thought long and hard about Hajar, after having discovered her story with us in Morgantown. She realized how little is said about this brave woman. "We have to take Hajar back and reclaim her, and realize that her suffering as a woman, as a single mother, as a wife, is so central to our tradition that every Muslim in this world is required to run back and forth during the hajj to re-create her footsteps! Yet when is the last time you heard a sermon talking about Hajar?" A possibility for the Muslim conference at Harvard: a Friday prayer led by a Muslim woman religious leader, who would also deliver the week's sermon. The woman: scholar Amina Wadud, who had been such an inspiration to me.

Almost two years earlier, I had stood in front of the Ka'bah without much awe. But an epiphany came to me as I started writing these words, the letters spilling onto the computer screen with each tap of my fingers, Shibli playing with Thomas the Train nearby.

I realized that standing in front of the Ka'bah had had the profound effect of showing me that I needed to stay true to a point of focus in my life. At that moment of the hajj, it was manifested upon my chest, gurgling and bright-eyed, in my son. It was at that moment, I realized, that I had made a commitment to dedicate my life to good. When I ascended to run in the path of Hajar, I shared my secret with the heavens. Over the next days of the pilgrimage, I came to understand more clearly what was important to me. Much of the hajj is one moment of seeming insignificance leading to the next; but the cumulative effect is transformative. It led me to reclaim my rights as a Muslim woman and seize control over my own identity.

I know the indomitable spirit of women in Islam—and women everywhere—will survive our lifetimes and triumph over more trials and tribulations. After all, in four thousand years, it is said, the well of Hajar has never run dry.

APPENDIX A

AN ISLAMIC BILL OF RIGHTS FOR WOMEN IN MOSQUES

1. Women have an Islamic right to enter a mosque.

2. Women have an Islamic right to enter through the main door.

3. Women have an Islamic right to visual and auditory access to the *musalla* (main sanctuary).

4. Women have an Islamic right to pray in the musalla without being separated by a barrier, including in the front and in mixed-gender congregational lines.

5. Women have an Islamic right to address any and all members of the congregation.

6. Women have an Islamic right to hold leadership positions, including positions as prayer leaders and as members of the board of directors and management committees.

7. Women have an Islamic right to be full participants in all congregational activities.

8. Women have an Islamic right to lead and participate in meetings, study sessions, and other community activities without being separated by a barrier.

9. Women have an Islamic right to be greeted and addressed cordially.

10. Women have an Islamic right to respectful treatment and exemption from gossip and slander.

APPENDIX B

AN ISLAMIC BILL OF RIGHTS FOR WOMEN IN THE BEDROOM

1. Women have an Islamic right to respectful and pleasurable sexual experience.

2. Women have an Islamic right to make independent decisions about their bodies, including the right to say no to sex.

3. Women have an Islamic right to make independent decisions about their partner, including the right to say no to a husband marrying a second wife.

4. Women have an Islamic right to make independent decisions about their choice of a partner.

5. Women have an Islamic right to make independent decisions about contraception and reproduction.

6. Women have an Islamic right to protection from physical, emotional, and sexual abuse.

7. Women have an Islamic right to sexual privacy.

8. Women have an Islamic right to exemption from criminalization or punishment for consensual adult sex.

9. Women have an Islamic right to exemption from gossip and slander.

10. Women have an Islamic right to sexual health care and sex education.

APPENDIX C

LETTERS:
THE SILENT MUSLIM MODERATE
MAJORITY SPEAKS

Tho' much is taken, much abides; and tho'
We are not now that strength which in old days
Moved earth and heaven, that which we are, we are
One equal temper of heroic hearts,
Made weak by time and fate, but strong in will
To strive, to seek, to find, and not to yield.

Alfred Lord Tennyson,
"Ulysses" (1842)

For a long time I believed that Muslim society was filled with hypocrisy. This may be true to some extent, but what I've since come to believe is that what seems like hypocrisy is actually contradiction. The actions of many Muslims seem to contradict their belief system. What I see is that they are actually in conflict with their cultural programming.

I have often stood physically alone in my life as a Muslim woman. But I have stood spiritually together with an ummah of honorable Muslim women, men, boys, and girls who, like my family, are trying to use Islam as a guide to living their days on this earth with love, honor, and respect toward others.

They are the silent moderate majority whose voices are so often lost in the thunderous assault of the hatred and violence perpetuated by certain Muslims in the name of Islam. They reached out to me as I walked on this unscripted path fulfilling my Qur'anic injunction to testify to the truths of injustice. These brave Muslims are my kin. And they did not abandon me or leave me alone in the parched desert. They stood by me, and they deserve the blessings that come with welcoming a person into the fold of Islam. With the love and understanding they expressed toward my son and me, they became for me the human expression of Islam's teachings of tolerance, compassion, and kindness.

I was blessed to be touched too by a wider global community of men and women who enact the universal principles of love and tolerance as the teachings of their churches, synagogues, temples, monasteries, and homes. They are a group of intelligent critical thinkers who do not judge the religion of Islam for the injustice, bigotry, and prejudice that have swept through it and who acknowledge that all of the world's religions and civilizations have been subject to these repressive forces. They separate the human dimension from the theological, and in reminding me that we are intertwined in our hopes, dreams, and visions, they have encouraged me.

Those who want to separate and divide the global community with hate find a twisted sense of belonging by abandoning the basic moral values of decency, respect, and kindness. I could fill these pages with the words of correspondents who have scolded, berated, and admonished me. Instead, I have chosen to share with you the voices that, it is my prayer, will gain ever more strength in this world.

Asra,

I was forwarded your police incident report which you and your father filed after the mistreatment at the mosque.

I was very shocked and angered at how both you and your father (long time members of the community) were treated after attending the mosque.

However, I was also not surprised. Just remember that ignorant people hold on to power the only way they can—through threats, violence and bullying. May Allah bless you and your father for standing up to the tyrants who hold onto the mosque as a place for themselves only. The only way they get away with it is because we let them. By taking a stand you and your father are not letting them get away with it any longer.

Just remember, they (your oppressors) do it out of fear and not out of what is right and just. From what I read, you handled it very well.

Please take care of yourself. I send this virtual hug to you to let you know that you are supported.

Your sister in the struggle.
Erum (The one from Calgary, Canada)

Dear Asra,

I am an architect working for a major pharmaceutical firm besides having my own independent but modest practice, mostly designing homes coupled with other miscellaneous projects. Like your father, I am also from Hyderabad and I expect I have much in common with him. I commend him for bringing up such a daughter as you and for the moral support and understanding he must be providing you. I am sure your mother must have also played a role in it. I am married to a wonderful lady, Fawzia. We have three very nice children, Faiz, Faraz and Farah.

I had suspected that you named your son Shibli for him to get inspiration from his ancestor, the great Urdu poet and author Maulana Shibli Nomani. I happened to read one of his books *Iblees ka Khutba-e-Sadarat* which is full of very refined humor and of high literary value. He was a person of great intellect and far ahead of his time.

Your fighting spirit and zeal must be applauded and what you have embarked upon is a great task—to get rid of the many ills our community is suffering from. It will take a great effort and energy and perseverance. You will also have to be very articulate. "*Samp mare aur lathi bhi na totae.*" "Destroy a snake, but don't break the stick."

I think we have found in you a person with the right qualities to lead us out of the darkness that prevails around us, into enlightenment. I also hope that your son grows into a man of integrity and courage and carries the torch you have lighted to far distances, inspired by you and Maulana Shibli Nomani. I wish you good luck.

> *Khuda Hafiz,*
> Naseer Ansari

I'm an Arabic gay who lives in an Arabic country but will go back to the UK soon. Tell me how can I be a pro-gay Muslim? Tell me how can I live my sexual life in a good way, but at the same time be a devoted Muslim.

Thanks.

I read your writings.

I think you're among the group of people who want the real Islam to return back, and I am supporting u and those people and praying that Allah is supporting u too.

Thank you.

Of course, my parents know nothing about my sexual identity, cause I'll face death if they know, unless, of course, I was in the UK.

Asra, this morning I read your article in the *Washington Post* before I went to Mass. I was struck with number one, your bravery and next the solidarity that we share.

As I worshipped this morning, I put myself in your shoes and wondered how I would react if women could not worship along with the men and children in my church. It didn't take long for me to realize that years ago, I would have been like many of the women of your faith and culture . . . but then, the peace of passivity would have been slowly eroded and in its place, frustration, anger and the continual questions of "Why"???

As a woman of 56 years, I am now coming into a place in my faith that requires courage and speaking out. Like Islam, Catholicism has been strongly influenced by the traditionalists and conservatives who want to "go back" to a time before Vatican II when we were so strict and inflexible . . . closed to other people of faith. In the 60's a courageous man, Pope John XXIII said, "Let us open the windows of the church and let the fresh air in." Many of us believe this was the Holy Spirit working through him. Now there is a movement within the church to destroy all that Vatican II achieved.

In May of this year, I went to Morocco with a few friends. At first, I felt so out of place and uncomfortable, but very soon, I realized that the Moroccan people had big hearts and lived their faith daily. Often I heard the word, "*Inshallah,*" and felt it resonate in my heart especially after learning it's meaning.

Asra, you are truly a woman of God, and I believe that He is inspiring you to work to free the restrictions of women in Islam, and I feel that your work will inspire courage in women of every faith to speak out. I see that all the major religions of the world are experiencing the "push" of conservatism in hopes of going back or, better yet, preventing the forthcoming freedoms that God has for all His people.

God bless you. . . . *Inshallah*
Linda Wright

From: Sarah Arif
Date: Friday, September 17, 2004 12:02 AM
Subject: Hats off to you.

Dear Asra,

This term of endearment comes natural here. I am proud of what you have accomplished and the courage you have shown in standing up for women's rights in Masjids.

My name is Sarah Sunita Arif, born in Hyderabad, India to a Brahmin family in the early 60s. . . . You can imagine the challenges, when I ran away from home to marry a Muslim boy when I barely turned 18 in the late 70s. . . . I accepted Islam with all my heart, although my motives could have been questioned at the time of my marriage.

Coming from a lineage of highly educated professors and doctors, I had a hard time with women being treated as second class citizens in Masjids. I have lived in Wichita Kansas since 1978 and yet to see a Masjid which will allow women to say their prayers along with men.

Some of us are planning to build a place of worship and a community center which will allow men and women praying side by side. I am hoping to gather moderate Muslims to join us in this task.

I wanted to thank you for showing me courage to join this group. I was a little hesitant till I read your stand on this issue. Hats off to you and my head is bowed!

Keep up the great work!

Sarah S. Arif

Dear Ms. Nomani,

Heard you on NPR this morning and read your homepage. You are a role model for women and men of all cultures. Please continue your struggle for equity as we all benefit from your courage, thoughts and actions.

Sincerely,
Hans Brauweiler, Ph.D.

ACKNOWLEDGMENTS

So many years all spent alone
Seeking a voice of my own

Elinor Jones (playwright),
A Voice of My Own

I am indebted to so many people for giving me the courage to find the voice that was my own. There is one moment that captures the importance of the simple acts of encouragement we can give each other.

It was the spring of 2004, and I was meeting a woman who was a pioneer in the American Muslim world. I wheeled Shibli in his stroller through the doors of the Ritz-Carlton in Tysons Corner, Virginia, to meet Sharifa Alkhateeb, a community activist born in Philadelphia to a Yemeni father and a Czech mother. I had first crossed paths with her after I broke the gender barrier at my mosque in Morgantown. Mrs. Alkhateeb was a well-respected Muslim community leader, and she didn't have to embrace a woman like me, dubbed radical by so many. But she did, quite literally, in the lobby of the Ritz. Together, we pushed Shibli in his stroller through the mall that adjoins the hotel. On our stroll, she offered me compassionate support. I was confiding in her the plan for a march on the mosque in Morgantown to affirm women's rights in the Muslim world. She offered encouragement and strategic guidance. "It is important what you are doing," she said, offering a blessing that gave me strength. The mother of three daughters, she died in October 2004 at the age of fifty-eight from pancreatic cancer. She will go down in history as an inspiration to American Muslim women like me to raise our voices. And I thank her and so many countless friends, colleagues, scholars, and readers who have given me encouragement to speak with a voice of my own.

For creating not only a book but also a vision, I am greatly appreciative of my publisher, Mark Tauber, and editor, Eric Brandt, at Harper San Francisco. They are grounded and yet visionary. For her tireless dedication to the spirit of my message, I am grateful to Miki Terasawa. I am thankful to Claudia Boutote for her enthusiasm. For their creativity and hard work, I am grateful to Terri Leonard, Kris Ashley, Jim Warner, Lisa Zuniga, Cindy Buck, and all of the dedicated staff at Harper San Francisco and Harper Collins Publishers. All of us have something important to say, it is true, and I am grateful to Kris Dahl, my literary agent at International Creative Management, for helping to make me blessed enough to share my thoughts with you, the reader. I also thank Liz Farrell and Carol Bruckner at ICM. And I am indebted to the work done by Jud Laghi, Mary Gollhofer, and Jonathan Baker behind the scenes.

My family and friends have generously shared with me countless hours of their commitment to the mission and vision. I thank my mother, Sajida Nomani, for her critical thinking, good sense, and important care of my son and me. I thank my father, Zafar Nomani, for his role in arranging our pilgrimage, his conscientious editing of my work, and his constant love. And I thank my brother, Mustafa Nomani, for his wise guidance and constant support. I am in awe of my friends Lynn Hoverman, Pam Norick, Ellyce Johnson, Vasia Deliyianni, and Rachel Kessler, who have helped me beyond the call of friendship. I am grateful to Mariane Pearl for her vision of my mission. For helping to bring my voice from a whisper to a roar, I am grateful to my coach, Ed Jacobs. This book and I were transformed because of his gentle coaxing and encouragement.

I thank Ron Shafer and Ken Wells, my former editors at the *Wall Street Journal*, and my countless friends from the *Wall Street Journal* for their friendship and guidance. I am indebted to Mohja Kahf, Patricia Dunn, Ahmed Nassef, Kecia Ali, Amy Leigh, and Alan Godlas for their editing. I have deep appreciation for the Islamic Society of North America and for ZamZam Tours and their guides for their attention to the details of my pilgrimage to Mecca. I am grateful to my local mosque in Morgantown and my local Muslim community for giving me the challenges that allowed me to grow. I thank the people and leaders of West Virginia, West Virginia University, and Morgantown, West Virginia, for their kind support, from Morgantown mayor Ron Justice to extraordinary citizens Sue Amos, Jim Friedberg, and Nellie Williams. For their prompt and courteous assistance, I thank the staff at the UPS Store, Office Depot, Superior Photo, and the West Virginia University Copy Center. And for a constant stream of coffee and a quiet corner without wireless Internet connection, I appreciate Mary Lewis and the kind wait staff at my local Morgantown diner, Eat 'n Park.

May all be as blessed as they have blessed me.

RECOMMENDED READING

BOOKS

Abdul-Ghafur, Saleemah. *Living Islam Out Loud: American Muslim Women Speak*. Boston: Beacon Press, 2005.

Abou El Fadl, Khaled. *The Great Theft: Wrestling Islam from the Extremists*. San Francisco: HarperSanFrancisco, 2005.

———. *The Place of Tolerance in Islam*. Boston: Beacon Press, 2002.

———. *Speaking in God's Name: Islamic Law, Authority, and Women*. Oxford: Oneworld Publications, 2001.

Ahmed, Laila. *Women and Gender in Islam: Historical Roots of a Modern Debate*. New Haven, Conn.: Yale University Press, 1992.

Algar, Hamid. *Wahhabism: A Critical Essay*. Oneonta, N.Y.: Islamic Publications International, 2002.

Ali, Samina. *Madras on Rainy Days*. New York: Farrar, Straus and Giroux, 2004.

Aslan, Reza. *No god but God*. New York: Random House, 2005.

Barlas, Asma. *"Believing Women" in Islam*. Austin: University of Texas Press, 2002.

Bewley, Aisha. *Muslim Women: A Biographical Dictionary*. London: Taha Publishers, 2004.

Hasan, Asma Gull. *Why I Am a Muslim*. New York: Farrar, Straus and Giroux, 2004.

Helminski, Camille Adams. *Women of Sufism: A Hidden Treasure*. Boston: Shambhala Publications, 2003.

Kahf, Mohja. *E-mails from Scheherazad*. Gainesville: University Press of Florida, 2003.

Knight, Michael Muhammad. *The Taqwacores*. Boston: Autonomedia, 2004.

Mernissi, Fatima. *Dreams of Trespass: Tales of a Harem Girlhood*. Reading, Mass.: Perseus Books, 1994.

Mernissi, Fatima, and Mary Jo Lakeland. *The Veil and the Male Elite: A Feminist Interpretation of Women's Rights in Islam.* Reading, Mass.: Perseus Books, 1992.

Murphy, Carlyle. *Passion for Islam.* New York: Scribner's, 2002.

Nafisi, Azar. *Reading Lolita in Tehran.* New York: Random House, 2003.

Rauf, Imam Feisal Abdul. *What's Right with Islam.* San Francisco: Harper-SanFrancisco, 2004.

Safi, Omid, ed. *Progressive Muslims: On Justice, Gender, and Pluralism.* Oxford: Oneworld Publications, 2003.

Viorst, Milton. *In the Shadow of the Prophet: The Struggle for the Soul of Islam.* New York: Anchor Books, 1998.

Wadud, Amina. *Qur'an and Woman: Rereading the Sacred Text from a Woman's Perspective.* Oxford: Oxford University Press, 1999.

Wolfe, Michael, ed. *Taking Back Islam: American Muslims Reclaim Their Faith.* New York: Rodale Press, 2002.

———. *The Hadj: An American's Pilgrimage to Mecca.* New York: Atlantic Monthly Press, 1993.

JOURNAL

Journal of Islamic Law and Culture. Edited by Aminah McCloud. Chicago: DePaul University.

READING ONLINE

Amanullah, Shahed, ed. AltMuslim, San Francisco, www.altmuslim.com.

Godlas, Alan, ed. The Islam Website, Athens, Ga., www.theislamwebsite.com.

Kazmi, Laila, ed. Jazbah, Seattle, www.jazbah.org.

Mazher, Uzma, ed. Crescent Life, St. Louis, www.crescentlife.com.

Nassef, Ahmed, ed. Muslim WakeUp!, Mt. Kisco, N.Y., www.muslimwakeup.com.

Rahman, Monis, ed. Naseeb, Karachi, Pakistan, www.naseeb.com/naseebvibes.

VIEWING

Kronemer, Alexander, and Michael Wolfe, producers. *Muhammad: Legacy of a Prophet.* Kikim Media and Unity Productions Foundation, 2002.

Nawaz, Zarqa, writer-director. *Me and the Mosque.* National Film Board of Canada, 2005.

Afterword

Muslim Women's Freedom Tour

With snowflakes tumbling around me, I walked up to the front door of my mosque and posted a simple treatise: "99 Precepts for Opening Hearts, Minds, and Doors in the Muslim World."

It was March 1, 2005—publication day for this book and the first day of Women's History Month, a good day to start the Muslim Women's Freedom Tour. I had brought the lessons from my pilgrimage to Mecca back home to Morgantown, and it was from this college town in the Appalachians that I could best share my vision for a new reality in our Muslim world.

As I taped up the precepts, I felt an exaltation I could not have predicted. I was standing up to the extremists in my mosque. When I was done at the front door, I went around back to the "women's entrance" and taped up a copy of the precepts there.

The precepts on the main door would come down quickly. Less than two hours later, the engineering professor and former mosque president who was presiding over my trial arrived for the early afternoon prayer. He studied the precepts, then raised his hand and ripped the poster off the front door. Tearing it in half and crumpling it in his hands, he paused only to shake his fist at a cameraman I had hired to document the day.

Tellingly, the copy I posted on the women's entrance would stay for days, undetected by the men, who would never bother to go around back.

Although they tried, the men of my mosque could not erase my precepts. If Martin Luther had the Wittenberg press to publish his precepts widely, I had the Internet. As I was posting the 99 Precepts on the door of the mosque, they were also being posted electronically on my website. Little did I realize then, however, that in posting the Precepts I had prepared myself to fire a shot about to be heard around the Muslim world.

The first stop on the Muslim Women's Freedom Tour was New York, where I planned to hear Dr. Amina Wadud's speak at Union Theological Seminary on "A Genderless God." Dr. Wadud, one of the first scholars I called when the men at my mosque told me that women must enter through the back door, had introduced me to the Islamic concept of *tawhid*, or the oneness of all beings that makes men and women spiritual equals. I called her now to ask, "Would you like to realize all of our dreams and lead a Friday prayer?"

Without hesitation, Dr. Wadud said yes. I could hardly believe it and immediately called Saleemah Abdul-Ghafur, the writer and activist who had led me and a few others in a mixed-gender prayer service on the banks of the Monongahela River with the Daughters of Hajar before our march of June 4, 2004, on the Morgantown mosque. "Are you sitting down?" I asked. "We can finally do it. We can have a woman-led Friday prayer!" Saleemah readily agreed to help me organize the day, and Ahmed Nassef, editor of Muslim WakeUp, agreed to be a sponsor and to advertise the prayer on his progressive Muslim website.

Our greatest challenge to the realization of this landmark woman-led prayer would turn out to be Muslims themselves. Muslim fundamentalists, both in

America and abroad, tried to stop it with the same intimidation tactics they use to try to silence anybody who opposes their hateful and sexist views. Imams attacked the woman-led prayer in their Friday sermons. A man e-mailed from New York: "Stop!!!!!!!!!!!!!" The New York venue I'd found for the prayer, Sundaram Tagore Gallery, run by a visionary descendant of the Nobel Laureate poet from India, received not kudos for advancing women's rights but hostile e-mails and phone calls, including a threat to "blow you up" and another to stage an "economic boycott." Ahmed Nassef was accused of being an agent of the West, not to mention the CIA.

I was packing my bags for New York when Sundaram Tagore called to say he had to back out after being alerted to a threat against Dr. Wadud posted in a Muslim chat room. He felt he had to protect his gallery staff. Though disappointed, I understood.

With less than a week to go, I searched desperately for a new venue. Call after call ended with discussions about "liability" and "insurance." The power of the extremists made me tremble. Then, just three days before our scheduled prayer, I called Mary Lyons at St. John the Divine, the Episcopal cathedral on Manhattan's West Side. She offered me a building on the church campus, the Synod House.

I couldn't believe my ears.

"Why are you helping us?" I asked.

"Why not?" she answered.

Indeed, why not. Why not allow women to reclaim their God-given rights within Islam? The Synod House turned out to be perfect, both spacious and, importantly for security, enclosed within a chain-link fence. I also made arrangements with the cathedral's security staff and the local NYPD precinct and hired a former NYPD officer as an armed bodyguard for Dr. Wadud.

Plus. Insights, Interviews, and More

On March 18, 2005, when about 125 Muslim women and men stood shoulder to shoulder behind a woman, we took back the faith from the 9/11 extremists and their followers and created a new reality in the Muslim world. As we gathered for prayer, a Muslim man greeted me with respect and kindness, and a woman breastfed her baby. When I stood before the congregation and introduced "The Islamic Bill of Rights for Women in the Mosque," I saw tears in the eyes of one of two women from my mosque in Morgantown. A woman, Sueyhla El-Attar, sang to the heavens the call to prayer that for too long in the Muslim world has been heard only through the voices of men. And I sang with Saleemah as she led us in chanting, "Oh light, oh light, oh light," thirty-three times. We were the light. We were rejecting the darkness.

But outside, Muslim protesters damned us to hell for daring to defy centuries of tradition. GENDER MIXED PRAYER TODAY. HELL FIRE TOMORROW, read one sign. Another reflected a stunning knowledge of my first book, *Tantrika: Traveling the Road of Divine Love*; it read: ASRA NOMANI, BEFORE YOU SPEAK ABOUT ISLAM, REPENT FROM YOUR TANTRIC SEX FANTASIES. The Internet buzzed with an e-mail carrying the subject line: "Stop Women from Leading Friday Mixed Prayers."

In the quiet of our sacred space, however, Dr. Wadud took her place in front of me, the first woman in the modern day to lead women and men in a public Friday prayer. As she said her final blessings, I turned to Saleemah and exclaimed: "We did it!" For days, I couldn't stop smiling. I decided that I would step forward and lead prayer myself at the next stop on the Muslim Women's Freedom Tour.

As a girl, I had loved reading the Qur'an and memorizing its lines in the original Arabic, as we had been taught. As I recited a chapter over and over again with my mother again as my tutor, I felt the pulse of my girl-hood return. I felt again the enthusiasm I had known for religion.

The second leg of the tour took me to Brandeis University outside Boston. There I led an early afternoon daily prayer on the lawn. There were just four of us present—myself, two young men who were friends of the Muslim punk novelist Michael Muhammad Knight, and Nakia Jackson, a young Bostonian who had asked me at the Islamic Society of North America's conference in Chicago how she could overcome her sense of aloneness at her mosque. Since then, she had become a symbol of courage, her struggles documented by PBS and chronicled by the *Boston Globe*.

Neither Nakia nor I had ever prayed before with a Muslim in purple hair and a leather bomber jacket. But this was our triumph, I believed: giving not only young women but young men safe spaces in which to practice their religion. That Friday, I asked Nakia to lead us in prayer, and she did, on the banks of the Charles River. Her sermon moved me with her allusions to God and light and was better than any I had heard in any mosque—a transforming breath of fresh air.

Back home in Morgantown, I called the local Unitarian Universalist Church, which readily agreed to let us use its space for a Friday afternoon prayer. This church had invited me to speak about my first book during a Sunday service—the first time I had stood in a pulpit—and I had broken my silence for the first time that day and spoken to a congregation of friends and neighbors about the challenges I was facing at my mosque.

But first I went to the regular mosque prayer in Morgantown. Not a single man said, "Salam." There was no joy around me. The sermon was about the "loyal wife," and the speaker encouraged the congregation to listen to a woman's advice—unless, of course, what she says is

Plus: Insights, Interviews, and More

wrong. For the first time, I walked out. I parted company with the congregation of the Morgantown mosque.

I had never felt better. In our new sacred space we would speak about light, our children's future, and the beauty of Islam. Several friends came for the first prayer service, as well as my mother, my father, and Shibli. I set out a fresh bouquet of roses and violets, moved the pulpit so that we would face east, and spread one of my mother's cotton tapestries on the ground. Then, after we had all gathered, I stood and delivered the sermon. "*Noora* is a feminine version of light," I told the congregation. "There are so many who are trying to perpetuate darkness in the name of Islam. By standing up for women's rights, justice, and tolerance within Islam, we are light." It was a profound and liberating moment. Afterward, our children blew soap bubbles to the heavens, and the sunlight piercing their ethereal creations seemed to engulf us in floating rainbows. I knew we had the power of the divine with us for we were standing up with love for Islam and this world.

My return to New York was marked, however, by a frightening confrontation. A reporter for the *New Jersey Star-Ledger* wanted to interview me about the struggle to win women's rights in our Muslim communities, and I thought there was no better place for us to meet than a mosque. I knew that the mosque on 96th Street, the Islamic Cultural Center of New York, had a "sisters' section" resembling an office cubicle behind the main hall, separated by a curtain. When we arrived, it was time for the late afternoon prayer. I ignored the women's section and prayed instead about fifteen feet behind the men in the main hall. A security guard reprimanded me: "Women in the back!" he yelled.

I said, "No. I have a right to pray in the main hall." I returned to my prayers. When I had finished, the security guard was waiting for me. "The imam. He wants to speak to you." I felt like a truant schoolgirl being called into the principal's office. The imam, Sheikh Omar Saleem Abu-Namous, stared at me from behind his desk, then reprimanded me for praying in the main hall. I stood by my right. He argued that the women had asked for the curtain. That was fine, but I didn't accept the curtain, and I was in America, where I could assert my rights. The conversation soon deteriorated as the imam mocked America.

He said, "America must submit to Islam, not Islam submit to America."

I responded, "America must submit to Islam?"

"Yes."

I thought of the electric ticker the mosque had at the corner of 96th Street above its gates, welcoming visitors. "Why don't we put that on the sign you have outside telling all of New York City that 'America must submit to Islam.'"

He continued to stun me. "If it does not submit now, there will come a time when it will submit in the hereafter."

"Wow. And how is that going to happen in the current day? How are we going to get America to submit to Islam, tell me? What is your idea?"

"I don't know. That's what America should do."

"And so how should it be expressed?"

"The fact that America has gone secular, this doesn't mean that it is right."

"So it should go Islamic?"

"Yes. Why not?"

"America should become an Islamic nation?"

"Yes, why not?"

"We should have sharia in America?"

"Yes."

"We should define American law by Islamic law?"

"Yes."

"We should require that all people in America live by Islamic law?"

The imam tried to interrupt: "Sister . . ."

I wanted an answer. "Tell me. Yes? Yes or no?"

"Yes."

"We should all impose the beautiful laws of sharia, such as adultery being punished by stoning, in America?"

"Yes."

"We should cut off the hands of those who steal in America?"

The imam flipped through the Qur'an to read the citation that the puritanical clerics use to justify amputation for thieves. I wanted a response. "Could you answer?"

"Yes."

"We should make sure that women do not wear jeans in America?"

"Yes, yes."

The imam proceeded to argue that women are to be blamed for wearing immodest clothing when they're raped and that I wasn't a "human being" because I didn't cover my hair in public. We had debated so long that it was time for the sunset prayer. "I want to pray," I told the imam as the reporter and I left his office and walked through the hall. "Pray," the imam said, "behind the curtain." I couldn't believe his obstinacy. But I was equally stubborn. "I will not pray behind the curtain." He responded, "Then you must leave." As the security guard and an angry man who had glared at me earlier in the imam's office stood menacingly by, I asked the imam, "So you will throw me out because I want to claim my rights as a Muslim woman?" "Yes," he said with a wide smile.

I went outside and unfurled a scarf onto the sidewalk, forced by the imam's unacceptable conditions to pray in the street.

My next stop: Washington, D.C. The Islamic Center of Washington is America's flagship mosque, sitting on Embassy Row. When I was in graduate school at American University, I had gone there once to pray. Once had been enough. I wasn't allowed to step inside the main hall, and a man pointed me toward a ridiculous annex where the floorboards creaked. For an annoying eternity, I listened to a crackle on the loudspeaker that was supposed to be the preacher's sermon. I left angry and disenfranchised and never returned until a Friday in April 2005, when I walked up to the front doors of the mosque with a new friend, Rafat Khan.

By then, I knew very well the tactics for managing women at mosques: steely eyes and firm fingers pointing to some door other than the front door. I also knew the tactics of civil disobedience. Rafat and I rejected the option of going downstairs to a cluttered and dingy basement room near the entrances to the bathrooms where women watched the Friday prayer on TV. "Thank you, but I'll be going in here," I said, moving confidently past the mosque bouncers and into the main hall. We were so early that only a few dozen men were present. The manager of the mosque had told us the day before that women couldn't pray in the main hall for the Friday prayer because there wasn't room for them. "Send a letter to the board of directors," he said when we protested. I stepped to the left and sat down against the back wall. Rafat sat to my left. "What do we do now?" she asked. "Now we pray," I said. She had been so afraid the night before that she almost backed out. I didn't want to pressure her, so I had reminded her, "It's all your choice." She showed up the next day. "We have to do it," she said.

The men started arriving to tell us the error of our ways. I told Rafat, "You don't have to argue now. Just be polite, but stay as long as you want to stay." Now was the time for certain action, not debate. A man to my right told me, "This is not Islam." I conveyed my certainty that it was indeed Islam. "You must convince the ummah," he said, waving to the room now filled with about three hundred men. I laughed. Social justice can't be subject to a popularity contest.

We sat at attention through the imam's sermon. He argued that men must respect women. It was a good speech. As the men started leaving at the end of prayer, the man who had been to my right said to me, "Thank you, sister. I learned something from you today." Another man made me smile when he said, "You're doing the right thing, sister." For that moment, we had changed the reality at the Islamic Center. We emerged into the sunshine triumphant.

That was not the feeling with which I would leave the Idriss Mosque in Seattle, where I had gone to give a book reading. When I took a space about fifteen feet behind the prayer leader in the front of the mosque, a man rushed over to exclaim: "We have a special place for women. Downstairs." He also gestured to a balcony with a screened wall. Although he seemed confused about where he wanted me to go, he clearly wanted me to leave the main hall. I told him, "I prefer to pray here." I cited the Fiqh Council of North America's ruling that it is not Islamic to require a woman to pray in a separate area. I cited the tradition of the Prophet in the seventh century that women are not required to pray behind a partition. Refusing to hear me, a dozen men harassed and surrounded me over the next hour, insisting that I leave the main hall. They had plenty of room to assemble for prayer fifteen feet in front of me, but they refused to hold the prayer. I felt so threatened that I pulled out a camera I had with me. They moved away from me when they saw it, but individuals kept coming back, speaking to me in a menacing way. One man threatened, "God will judge you on your judgment day." I said, "I am doing nothing wrong. Pray."

I was reminded that this battle isn't divided according to gender lines. A woman yelled through the screen separating the balcony from the main hall, "This is not Islam. Come here!"

Finally, the men threatened to call the police. I was ready to be arrested, but a friend accompanying me wasn't. I left angry and annoyed. "Thank you for your hospitality," I said, unable to resist a sarcastic comment. "Don't ever come here again! You're banned from here," a man bellowed. As I walked away, one of the men followed, still hostile and intimidating. That night I opened my book and read: "I am coming to you from the frontlines in the struggle for the soul of Islam."

Going to San Francisco was like returning to my youth. When I started there at the *Wall Street Journal* as an intern in the spring of 1988, it never crossed my mind to go to a mosque. This time I decided to visit the Islamic Society of San Francisco in the heart of the Tenderloin District.

Walking past drunks, drug addicts, and the homeless on the sidewalks, I turned up the stairs into a massive loft. I went behind a thick wall and saw half a dozen women sitting in a large space, some of them napping. I sat down in front of a portable TV set. *Ridiculous*, I thought to myself. I got up, walked around the wall, and took a seat nearly to the back of the hall but not completely. The area was massive, and the closest men were many feet away. Immediately, however, a burly man approached me. "Sister, you must sit in the back." I responded with what had become my mantra: "No, thank you. I'm happy praying here."

"Do not cause *fitna*," he said, using the catchall term of intimidation meant to silence protest. "I know you are here to pursue some agenda."

Plus: Insights, Interviews, and More

"Please," I responded. For the first time, I lost my patience. It's so simple for people like him to dismiss others with the claim that they're pursuing a nefarious agenda. But then something remarkable and unprecedented happened: a man stepped forward to protect my rights. He told the man that the mosque president, Souleiman Ghali, had said women could pray in the main hall. I couldn't believe my ears. Here was a sign of leadership in our Muslim world. Nevertheless, the protesting man persisted, so I resorted to my defense tactic: I started praying. "Interrupt me if you will, but I will testify against you on your judgment day." He retreated.

Then another remarkable thing happened: the president's sermon was a testimonial to a woman convert who had been a tireless volunteer. Afterward, he welcomed me, saying, "I want more women to come out from behind the wall." I thanked him for his vision.

In southern California, I had planned a visit with Dr. Khaled Abou El Fadl, the UCLA professor of law whom I'd come to consider the pope of tolerant Islam. His kind assistant, Naheed, greeted me with an embrace and the question, "What's going on? They're organizing a town hall meeting on woman-led prayer. The professor asked for you to be there, and they said you're too controversial. The professor thinks that is so ridiculous."

I had faced off against scary men in Seattle and stared down a burly man in San Francisco. But it was rejection by people I thought of as kindred spirits that was threatening to break me. Some people had begun to perceive me as a liability. The criticism had been fierce after the New York prayer, which I was accused of organizing just to sell books. That stung, because I had been active in this struggle long before my book came out. I hadn't confided my hurt to anyone until I spoke to Naheed. "I was afraid to come here," I admitted. "I was afraid that Dr. Abou El Fadl and all of you would also turn on me."

"No way," Naheed said, gently putting her arm around my shoulders.

When I greeted Dr. Abou El Fadl in his library, he moved me to tears with his support and understanding. "You are a writer. You write books. You came here to seek knowledge. I helped you. You came to a recognition for which you took action. Most people do nothing with the knowledge they get. I have nothing but admiration for you."

I gave him a gift: the new "Islamic Bill of Rights for Women in the Mosque," complete with the rights he had made me understand we could assert to be prayer leaders to both men and women and to stand in mixed-gender lines in the front of mosques.

I knew that I had to go to the Islamic Center of Southern California, the flagship mosque of Los Angeles. It was supposed to be one of the most liberal mosques in America and had women among its leaders. But I had also heard that women there were segregated in peculiar ways.

Arriving just before the Friday prayer, I walked into a large reception area lined with cubbyholes for shoes. To the right was a room separated from the reception area by tinted glass—the "women's section." To the left of the reception area was another room with the pulpit—the main prayer hall. I stepped into the women's section and immediately felt disconnected from the main hall, with no clear visual and auditory access to the imam.

I stepped into the main hall, which was not yet filled, found a space at the extreme right side of the room, and did a prayer. I then sat down for *zikr*, or remembrance of God, to wait for the sermon and prayer to begin. A man told me: "The sisters' section is in the back. You must leave." I said, "I will be praying here." Other men came

Plus: **Insights, Interviews, and More**

and told me to leave. I politely told them that it is Islamically legal for women to pray in the same space as men. One man said, "It is *haram*." I said, "It is *halal*." Another said," Let us remove her. Call the women to lift her and remove her."

I continued my *zikr*, counting each recitation on the digits of my fingers. A woman arrived and told me, "Sister, your place is in the back." I said, "It is Islamically legal for me to pray here." She told me that she had three points to make. "People come to Friday prayer for peace," she said. I said, "Then let us just pray." "Women and men have to be separated," she said. I noted that women and men were not separated by partitions at the time of the Prophet in the seventh century. And finally she said, "I am your elder. I am like your mother. You have to listen to me." I told her, "My mother supports me." The woman grabbed my elbow and tried to lift me up and remove me. Although older, she was much larger than me, and physically aggressive. She said, "Let's discuss this over there," nodding her head behind me. I looked over. She was gesturing to a fire exit door. I declined, and she kept prodding and pushing me. I protested her physical harassment: "Please do not touch me." I shook my elbow from her hold, stood up, and said, "I am going to pray. I would hope you would not interrupt me." She did interrupt me, pushing against my shoulder to force me out the fire exit door. I sat down immediately and returned to my *zikr*. I looked down and saw that my hands were trembling. I feared for my physical safety.

Then a man arrived and said, "There is a TV camera here. Let her pray. Call the women to join her." A group of about six women joined me for prayer, and the mosque officials surrounded us with a security rope. About 150 men prayed to our left. A man whispered to me, "Sister, you're doing the right thing. You have a right to pray here." I thanked him. He looked away.

Afterward, the mosque leader, Dr. Maher Hathout, called on members of the congregation to sign a petition against alleged desecration of the Qur'an by the U.S. military at Guantanamo Bay. He stood in the pulpit to rally support for protecting an object—a symbolically important object, but nonetheless an object—but failed to stop the physical harassment in his own mosque of a human being. He also spoke about the mosque's rights as a private organization with 501-C-3 tax-exempt nonprofit status. The irony of using America's tax laws to protect the right to gender discrimination made me laugh.

I made it back home from a road of much trouble but great victory. During my travels with the Freedom Tour, my son, Shibli, had been a real trouper, so to reward him I took him to Hershey Park, a few hours from home. After a weekend of playing games and riding every train in the park, I checked my cell phone messages as we left. "Hello, Asra," a man's voice said, slipping straight into Urdu. "Bitch's daughter. If you want to stay alive, then keep your mouth shut. I am going to slaughter you *halal* style." The Islamic technique for killing animals is to say a Muslim prayer, slice the animal's jugular vein, and behead it. That was how Danny's murderers killed him.

"Bitch's daughter," the message continued. "I am going to slaughter your mother and father *halal* style. Slaughter. Pray as many prayers as you want. For now, keep your mouth shut. Do it. Otherwise I will slaughter you *halal* style. Think hard before you do anything. Think hard. I know your address. I know your mother and father. If you want to keep them alive, then keep your mouth shut. *Qhuda hafiz*," an Urdu salutation that means, ironically, "May God protect you." I was stunned.

I looked behind me. Shibli had fallen asleep. My mother called at just that moment, and I told her, "I got a weird phone call." She said, "We did too." The same man had called her, and she had hung up on him.

Knowing this was serious, I called 411 and asked the operator for the city that matched the caller's area code: Chico, California. The Chico police dispatcher couldn't find the number in her records. I pulled into the McDonald's at the intersection of highway 39 and I–81 in Linglestown, Pennsylvania, ducking inside. Over the next hour, I reported the case to the Monongalia County sheriff's office. The deputy there knew my case from a memo that had been sent around after my publisher received a phone call warning that I was in danger. I also talked with a special agent for the FBI's Joint Terrorism Task Force.

A few days later, FBI agents told me what they had learned: the number was a cell phone registered to a young Laotian American man with ties to gangs, drugs, crime, and armed violence. "If a man with a tattoo of a three-headed elephant rings your doorbell, don't answer it," one of the agents warned. I suddenly recognized what I had started to see after the woman-led prayer earlier in the spring: there were some who would turn to violence in this war within Islam. In the *New York Times*, reporter Andrea Elliott identified the young men who stood outside our march as members of an extremist Queens Muslim group who also opposed America. We are smoking out the extremists when we challenge male control.

In Morgantown, my own spiritual community was evolving. Before a community meeting organized by young reformers at the mosque, the professor who had made sure I knew I wasn't welcome at community dinners the past Ramadan e-mailed to tell me that had been a committee decision, not his own decision. He understood that a Muslim should ask for forgiveness before embarking on a journey. Did he want my forgiveness?

Was he apologizing? I asked him these questions when he approached me before the meeting. I wanted to know if he understood how painful his message had been to me. "I apologize," he said. I accepted his apology.

I still had to sit on trial to be banned from the mosque. I had been told that a panel recommended that I *not* be banned. The board was required to act on the recommendation, but even after repeated requests for a decision, I got the typical response: silence. This experience has underscored what I see as an important goal: this era of pathological passive-aggressive leadership must end if the Muslim world can ever hope to progress. As ordinary Muslims, we must hold our leaders accountable.

My hope for the Muslim world took a hit from the explosion of bombs in London on March 7, 2005. The killing of innocent people in London sent a simple message to Muslims: we are failing in our communities against the seductive forces of the extremists, especially with our youth. If I had any doubts about this, they were laid to rest when I received my third death threat, which I traced back to Seattle. It was from a writer with the e-mail address "Kalaamulaah Ali," *kalaamulaah* meaning "black cleric" in Urdu. The subject line read: "stop evil act." The message, for all its typos and bad grammar, was simple: "it is not freedom what you call freedom but it is evil act islam was long time and we never hear women can pray next to men or Women leads mixed-gender prayer before you but if what you say is right we will not be muslim at all but we know what you say is what Bush tell you to do if you are jew or evil we will find you one day and cut your head off it is easy for us to do that if you try to keep your evil act it is our job to kill who like to say bad think our islam if you like to get out home and be save shut your mouth or we send you to hell fire we have your picture."

I knew what I had to do: I called the FBI.

By June 2005, we had made women's issues and the reform of our communities a priority for even the national
Muslim organizations, which issued a twenty-eight-page
document, "Women-Friendly Mosques and Community
Centers: Working Together to Reclaim Our Heritage." We
haven't won the war, but with this document, we have
won a battle. The report represents a huge victory for
women's rights in mosques and affirms every single right
in the "Islamic Bill of Rights for Women in the Mosque"
except for two I had added: the right of women to be
prayer leaders to both men and women and the right of
women to stand in the front row with men. No longer can
a man stand up in front of a woman and claim that Islam
protects his right to order her into the far corners of the
mosque. No longer can a man stand up and declare a
woman's voice is not to be heard in the mosque. No
longer can a man stand up and declare that a woman has
to be deaf and dumb if she knows her rightful place in
Islam. In recognizing women's rights, the American Muslim community has acted as the shining light I have
known it can be to the rest of the world, rejecting antiquated traditions and affirming Islam. The next step: realizing these rights.

99 Precepts for Opening Hearts, Minds, and Doors in the Muslim World

1. The Loving One: Live with an open heart to others.

2. The Only One: We are all part of one global community.

3. The One: All people—women and men, people of
   all faiths, cultures, and identities—are created and
   exist as equals.

4. The Self-Sufficient: All people—women and men,
   people of all faiths, cultures, and identities—have a
   right to self-determination.

5. The Creator of Good: All people have a human right to happiness.

6. The First: A fundamental goal of religion is to inspire in us the best of human behavior.

7. The Preserver: Religion isn't meant to destroy people.

8. The One Who Gives Clemency: We aren't meant to destroy people.

9. The Absolute Ruler: We are not rulers over each other.

10. The Owner of All: No individual or group of individuals may treat any of us as property.

11. The Mighty: Spirituality goes far deeper than mere adherence to rituals.

12. The Appraiser: We are the sum of our small deeds of kindness for others.

13. The Inspirer of Faith: It is not for human beings to judge who is faithful and who is not.

14. The One with Special Mercy: Humanity and God are best served by separating the "sin" from the "sinner."

15. The Finder: Virtue doesn't come with wealth.

16. The Supreme One: All people are created with an inner nature that seeks divine nature and is disposed toward virtue.

17. The Doer of Good: Thus, live virtuously.

18. The Greatest: Have the courage to take risks.

19. The Possessor of All Strength: Have the courage to stand up for your beliefs, for truth, and for justice even when they collide with the status quo.

20. The One Who Honors: Respect one another.

21. The Magnificent: Glorify one another with kind words, not harsh words.

22. The Forgiver: Forgive one another, and ourselves, with compassion.

23. The All-Compassionate: Be compassionate with one another.

24. The Compeller: Love the soul even when we don't love the "sin."

25. The All-Merciful: Be motivated by love of God, not fear of God.

26. The Supreme in Greatness: Be kind, respectful, and considerate with one another.

27. The One Who Rewards Thankfulness: Appreciate the freedoms you enjoy.

28. The Accounter: Know that we are all accountable for how we treat one another.

29. The Gatherer: Know that anyone you wrong will testify against you on your judgment day.

30. The Expander: Be friends to one another.

31. The Exalter: Win the greatest struggle—the struggle of the soul, *jihad bil nafs*—to good.

32. The Highest: Rise to the highest principles of Islam's benevolent teachings.

33. The Giver of All: Rise to the highest values of human existence, not the lowest common denominator.

34. The One Who Opens: Live with an open mind.

35. The One Who Enriches: The Qur'an enjoins us to enrich ourselves and our communities with knowledge.

36. The Subtle One: Islam is not practiced in a monolithic way.

37. The All-Forgiving: We allow ourselves to be more positively transformed if we accept rather than despise our dark side.

38. The Maker of Beauty: Islam can be a religion of joy.

39. The Maker of Order: In any society governed by oppression and senseless rules, there will be rebellion, whether expressed publicly or in private.

40. The Guide to Repentance: Evil is social injustice, discrimination, prideful rigidity, bigotry, and intolerance.

41. The Nourisher: We were all created with the right to make our own decisions about our lives, our minds, our bodies, and our futures.

42. The One Who Withholds: Certain traditions and ideologies betray Islam as a religion of peace, tolerance, and justice.

43. The Creator of the Harmful: Repression creates fears that are manifested in dysfunctional ways.

44. The Generous: Women possess the same human rights as men.

45. The All-Comprehending: Chastity and modesty are not the sole measure of a woman's worth.

46. The Last: Puritanical repression of sexuality and issues of sexuality is self-defeating and creates a hypersexual society.

47. The Seer of All: The false dichotomy between the private world and the public world leads us to avoid being completely honest about issues of sexuality.

48. The Majestic One: The Qur'an tells us: There is no compulsion in religion.

49. The All-Aware: The Qur'an enjoins us: Exhort one another to truth.

50. The Knower of All: Thus, seek knowledge.

51. The All-Powerful: Do not put any barriers in front of any person's pursuit of knowledge.

52. The Ever-Living One: Reject ignorance, isolation, and hatred.

53. The Truth: Live truthfully.

54. The Praised One: Praise worthy aspiration, not destruction.

55. The Manifest One: Be the leader you want to see in the world even though you lack position, rank, or title.

56. The Perfectly Wise: Lead with wisdom.

57. The Originator: Open the doors of *ijtihad* (critical thinking) based on *istihsan* (equity) and *istihsal* (the needs of the community).

58. The One Who Is Holy: Honor and respect the voices and rights of all people.

59. The Sustainer: Empower each other, particularly women, to be self-sustaining.

60. The Governor: Do not allow anyone to unleash a vigilante force on any man, woman, or child.

61. The Hearer of All: Be honest about issues of sexuality in our communities.

62. The Expeditor: Lift repression.

63. The Guardian: Reject a sexual double standard for men and women.

64. The Restorer: Reform our communities to reject bigoted, sexist, and intolerant practices.

65. The Righteous Teacher: Question defective doctrine from a perspective based on the Qur'an, the traditions of the Prophet, and *ijtihad*.

66. The One Who Resurrects: Know that we all will face a reckoning for our deeds.

67. The Guide: We must open the doors of Islam to all.

68. The Creator of All Power: We are in a struggle of historic proportions for the way Islam expresses itself in the world.

69. The Mighty: The Qur'an is clear: Stand out firmly for justice, as witnesses to God, even if it may be against yourselves, or your parents, or your kin.

70. The Satisfier of All Needs: Political expediency does not override our morally compelled duty to tell the truth.

71. The Responder to Prayer: Spiritual activism is a noble pursuit.

72. The One Who Humiliates: Sexism, stereotypes, and intolerance are the common denominators of all extremism.

73. The Giver of Life: We cannot accept murder in the name of Islam.

74. The Inheritor of All: Racism, sexism, and hatred are unacceptable in God's world.

75. The Taker of Life: Dogmatism and intolerance lead to violence.

76. The One Who Abases: Making women invisible is a defining feature of violent societies.

77. The Just: Women and men are spiritual and physical equals.

78. The Equitable One: Women's rights are equal to men's rights.

79. The Witness: Nothing we do is without a witness.

80. The One Who Prevents Harm: Rejecting injustice is more important than protecting honor.

81. The Delayer: Honor can be the worst expression of ego.

82. The Judge: Justice is not what the majority believes is right.

83. The Forbearing One: We are not judges upon each other.

84. The Ruler of Majesty and Bounty: If change will come tomorrow, we should not wait but should create it today.

85. The Trustee: Thus, know women have an intrinsic right to be leaders in all capacities in our Muslim world, including as prayer leaders or *imams*.

86. The Creator: Reach inside to create the change you want to see in the world.

87. The Forceful One: Stand strong for justice.

88. The One Who Subdues: Stand up to extremists and all forms of extremism.

89. The Self-Existing One: Break the silence sheltering injustice and intolerance.

90. The Originator: Create a new reality.

91. The Glorious: Stand up to the forces of darkness.

92. The Watchful One: Question the source of hate in order to dismantle it.

93. The Protector: Respect women's equal rights and human dignity, from the mosque and the public square to the workplace and the bedroom.

94. The Avenger: Use principles of social justice to define our communities.

95. The Everlasting: Stand up to create an everlasting Muslim world that will enrich our global society.

96. The Patient One: Exercise patience as a virtue, not as an excuse.

97. The Source of Peace: Live peacefully with others.

98. The Light: Create cities of light to overpower the darkness in our Muslim world.

99. The Hidden One: Ultimately our choice is only one: We must create communities with open hearts, open minds, and open doors to all.